# SpringerBriefs in Applied Sciences and Technology

More information about this series at http://www.springer.com/series/8884

Wilson Acchar · Sheyla K.J. Marques

# Ecological Soil-Cement Bricks from Waste Materials

 Springer

Wilson Acchar
Department of Physics
Federal Univ of Rio Grande do Norte
Campus Universitario
Natal, Rio Grande do Norte
Brazil

Sheyla K.J. Marques
Campus Palmeira dos Índios
Instituto Federal de Alagoas (IFAL)
Palmeira dos Índios, Alagoas
Brazil

ISSN 2191-530X          ISSN 2191-5318   (electronic)
SpringerBriefs in Applied Sciences and Technology
ISBN 978-3-319-28918-2          ISBN 978-3-319-28920-5   (eBook)
DOI 10.1007/978-3-319-28920-5

Library of Congress Control Number: 2016932512

Printed on acid-free paper

This Springer is published by Springer Nature
The registered company is Springer International Publishing AG Switzerland

# Foreword

Among the global challenges listed in the Millenium Project, sustainable development has been a growing concern with particular emphasis to its effect on climate change. Due to its continental dimensions, it is expected that Brazil plays a significant role in the agenda for environmental preservation, and a good example of successful participation can be found in the proportion of renewables in our energetic matrix, around 40 % in 2014, compared to less than 9 % from the OECD countries. The two main sources for this expressive result are the biomass derived from sugar cane and hydroelectricity. Another important action to reduce the environmental damage is the reduction of raw material use in large-scale enterprises, such as building and road constructions.

The present work by Prof. Wilson Acchar addresses this important issue with a detailed description of the benefits promoted by the incorporation of two available waste materials, gravel from oil drilling waste and sugarcane bagasse ash into soil-cement bricks, a component largely used in the building of house units, a major area of Civil Engineering. Supported by an extensive experimental research, the author shows that the properties of the new materials displays better properties than the soil-cement mixture traditionally employed in the fabrication of bricks, exceeding the requirements of Brazilian standards.

The structure of the book offers an interesting and adequate sequence of information, starting with a description of historical aspects about the introduction of soil-cement bricks in ancient times and its evolution as a reliable and affordable building component; this includes the attractive possibility it offers to incorporate waste material generated by many industrial activities, contributing to reduce the environmental impact. Next, the origin and characteristics of the two waste materials studied, soil drilling waste from the petroleum industry and sugar cane bagasse ash are described. The next two chapters present, for each type of waste, the complete methodology and experimental procedure developed, including the preparation of the soil-cement bricks and the resulting superior mechanical and water absorption properties obtained, which support the conclusions of the author proposing their generalized use in house-building programs. The book is certainly a welcome and relevant contribution to those fields where the use of soil-cement bricks is recommended.

Fernando Cosme Rizzo Assunção
Director of National Institute of Technology (INT)
Rio de Janeiro-RJ, Brazil

# Contents

# Chapter 1
# Introduction

**Abstract** This chapter introduces the soil-cement brick and wastes to be incorporated in ceramic formulations.

**Keywords** Soil-cement brick · Soil drilling waste · Sugarcane bagasse ash

Among various types of buildings, the execution of housing units is a major area of Civil Engineering, whereas there is a housing deficit in Brazil, the solution proposed by the civil construction sector and the government is to intensify the number of buildings to meet the demand of low-income families. As a result of the rapid population growth, the number of works is increasing (Souza 2000). In the construction of affordable housing, the use of soil-cement bricks, also known as green bricks, is widespread due to advantage such as rapid manufacturing in the own construction site, hand labor to operate equipment does not need to be specialized and may be made by the community, as well as, good quality and regularity in the final appearance of the product, providing lower consumption of laying mortar and coatings.

Thus, works are concluded in shorter deadlines promoting a quite satisfactory cost-benefit ratio. Soil-cement is the material resulting from the homogeneous, compacted and cured mixture of soil, cement and water in suitable proportions. The product resulting from this process is a material with good resistance to compression, good impermeability index, low volumetric shrinkage index, and good durability. In terms of soil-cement bricks, the largest amount of raw material is directed to the soil. Soils are complex mixtures of inorganic materials and partially decomposed organic waste, whose composition widely differs from location to location, not only in quantity but also qualitatively. For this reason, it is widely used as raw material in the civil construction sector, including in the manufacture of soil-cement bricks.

In addition to bricks, soil-cement applications requiring lower construction costs are: containment of slopes, highway containment barriers (retaining bags), road subfloors on highways (compacted soil-cement), and carriageway on airports (compacted soil-cement). The incorporation of waste is one of many conditions to

© The Author(s) 2016
W. Acchar and S.K.J. Marques, *Ecological Soil-Cement Bricks from Waste Materials*, SpringerBriefs in Applied Sciences and Technology, DOI 10.1007/978-3-319-28920-5_1

increase the economic viability, since their generation is inevitable. The potential benefits of recycling to society are preservation of natural resources, energy saving, reducing the volume of landfill sites, pollution reduction, job creation, reduction of environmental control cost by industries, increased durability, among others (John 2000). The inclusion of reinforcing elements to soil to improve its properties was already known to people of antiquity. This can be confirmed by some buildings that still exist today such as the walls of Ziggurat of Agar Quf in Mesopotamia (1400 BC), built using layers interspersed with soil and plant roots. The Incas used llama wool mixed with soil in the construction of roads resistant to time (Palmeira 1992).

Oil production in Brazil increased significantly in recent decades due to continuous technological advances in drilling and production. Drilling fluids used in the drilling of oil wells are complex mixtures of solids, liquids, chemicals, and sometimes even gasses. The main purpose of the drilling fluid is to carry drilled rock fragments to the surface; cool and lubricate the drill and hydraulically support the oil well walls. Drill gravels are mixtures of small rock fragments impregnated with the fluid used to lubricate and cool the drill during drilling (Leonard and Stegemann 2010). This type of oil well drilling waste is disposed in landfills without prior treatment and constitutes a major environmental problem.

The sugar-alcohol sector has greatly expanded in recent years, mainly driven by the advent of renewable energy. In Brazil, the production of sugarcane has advanced over areas for extensive cattle ranching and other important crops such as soybeans, corn, and orange, both for the generation of electricity from burning sugarcane bagasse (cogeneration or combined heat and power—CHP) and use of ethanol as fuel for motor vehicles. Although the cogeneration process releases carbon dioxide ($CO_2$) into the atmosphere, the amount of emissions is significantly lower compared to other energy sources such as oil (Freitas 2005). However, production of sugar, ethanol, and energy from sugarcane can cause other environmental problems such as the generation of waste: straw, straw ash—in the case of manual harvest, bagasse, and bagasse ash.

Faced with this global panorama of growing sugar-ethanol industry combined with sustainable development, it is of paramount importance to carry out studies aimed at the use of this waste, turning them into commercially valuable products. The incorporation of waste into soil-cement bricks is already a reality in scientific research and has brought positive results, both in terms of technological characteristics as in terms of environmental preservation. Thus, in this context, the use of these wastes in the civil construction sector means that both the financial cost savings as the perspective of sustainable industrial growth, resulting in better quality of life conditions for an economic portion of society. This book presents results obtained by the incorporation of soil drilling waste from the oil industry (SDW) and/or sugarcane bagasse ash (SBA) into soil-cement bricks. Experimental results have proven the efficiency and high potential of using waste from oil well drilling and sugarcane bagasse ash to produce waste-cement brick with higher mechanical strength and lower water absorption compared with conventional soil-cement bricks.

# References

Freitas ES (2005) Characterization bagasse ash from sugar cane of Goytacazes Campos municipality for use in construction. Dissertation (Master in Civil Engineering), Campos dos Goytacazes, State University of North Fluminense UENF, RJ, p 81

John VM (2000) Recycling of waste in construction: Contribution to research and development methodology. Thesis (Habilitation), Department of Construction Engineering, Polytechnic School of the University of São Paulo, England, p 113

Leonard SA, Stegemann JA (2010) Stabilization/solidification of petroleum drill cuttings. J Hazard Mater 174:463–472

Palmeira EM (1992) Geosynthetics: types and evolution in recent years. Seminar on geosynthetics in geotechnical applications, Brasilia, pp 1–20 (cited)

Souza PABF (2000) Recycling as a competitive strategy for the Construction industry. Brazilian Congress of Environment, Fortaleza

References

# Chapter 2
# Soil-Cement Bricks

**Abstract** This chapter describes the historical aspects, properties, and potential of soil-cement bricks.

**Keywords** Soil-cement bricks · Cement

## 2.1 Portland Cement

### 2.1.1 History

Portland cement is a fine powder presenting agglomerating, agglutinating, or binding properties that when under water, hardens and no longer decomposes when exposed to water again. Portland cement is composed of clinker and additions, and the former is its main component, which is present in all cements. Additions vary from one type of cement to another (ABCP 2002). The word CEMENT originates from the Latin word CAEMENTU, which in the old Rome designated a kind of natural stone that did not square. Cement dates back about 4500 years. The imposing monuments of ancient Egypt already used an alloy composed of a mixture of calcined gypsum. The great Greek and Roman works such as the Pantheon and the Colosseum were built using volcanic soils from the Greek island of Santorino or the surroundings of Italian town Pozzuoli, which had hardening properties under the action of water (ABCP 2000).The major step in the development of cement was taken in 1756 by Englishman John Smeaton, who managed to obtain a high-strength product through calcination of soft and clayey limestone. In 1818, Frenchman Vicat obtained results similar to those of Smeaton by mixing clayey and limestone components. He is considered the inventor of artificial cement. In 1824, English builder Joseph Aspdin burned together limestone and clay, turning them into a fine powder. He realized that he had obtained a mixture which, after drying, became as hard as the stones used in construction. In Brazil, the first attempt to apply knowledge related to the manufacture of Portland cement apparently occurred in 1888, when Commander Antônio Proost Rodovalho built up a factory in Santo

© The Author(s) 2016
W. Acchar and S.K.J. Marques, *Ecological Soil-Cement Bricks from Waste Materials*, SpringerBriefs in Applied Sciences and Technology, DOI 10.1007/978-3-319-28920-5_2

Antônio farm of his property located in Sorocaba, SP, Brazil. Subsequently, several sporadic initiatives to manufacture Portland cement were developed. A small production facility was operated for three months in 1892 in the island of Tiriri, Paraiba. The Rodovalho plant operated in 1897–1904, returning to work in 1907 and definitively extinguished in 1918. In Cachoeiro do Itapemirim, the government of the state of Espírito Santo founded a factory in 1912 that operated until 1924, resuming operation in 1936, after modernization. All these initiatives were only mere attempts that culminated in 1924 with the establishment by the Brazilian Company of Portland Cement in Perus, state of São Paulo, whose construction can be considered a hallmark of the Brazilian cement industry. The first tons were produced and placed on the market in 1926. Until then, cement consumption in the country depended exclusively on imported products. Domestic production gradually increased with the implementation of new plants and the share of imported products fluctuated during the following decades until almost disappearing in current days.

## 2.1.2   Manufacture

The raw material is extracted from mines by usual processes for the exploitation of mineral deposits. Limestone may have high hardness, requiring the use of explosives followed by crushing, or low hardness requiring only the use of disintegrators to be reduced to the size of maximum particle diameter of 1 cm. Clays containing silicates, alumina, and iron oxide are usually capable of being directly mixed with limestone. Limestone and clays, in predetermined proportions, are sent to the grinding mill (ball, bar, and roller mill), where the intimate mixing of raw materials occurs and at the same time, they are turned into powder to reduce their particle size diameter to 0.050 mm, on average. The determination of the percentage of each raw material in the raw mixture essentially depends on the chemical composition of raw materials and the desired composition to be obtained for Portland cement when the manufacturing process is completed. During the manufacturing process, the raw material and the crude mixture are chemically analyzed numerous times at intervals of 1 h and sometimes every half hour, and according to the test results, the laboratory indicates the percentages of each raw material that should compose the crude mixture. The properly dosed raw material and reduced to a very fine powder after grinding should have its homogeneity ensuring the best possible way. Once pulverized and properly mixed in suitable dose, the raw material undergoes heat treatment which can be seen in Table 2.1.

In the furnace, as a result of treatment, the raw material is transformed into clinker. At the outlet, the material is presented in the form of balls of maximum diameter ranging from 1 to 3 cm. The balls that form the clinker leave the furnace at temperature of about 1200 to 1300 °C, as there is a temperature lowering in the final stage, still within the furnace. The clinker leaves the furnace and enters the cooling equipment, which can be of various types. Its purpose is to reduce temperature,

**Table 2.1** Heat treatment of the raw material after dosing

| Temperature | Process |
| --- | --- |
| Up to 100 °C | Evaporation of free water |
| 500 °C and above | Dehydroxylation of clay minerals |
| 900 °C and above | Crystallization of decomposed clay-mineral compounds |
| 900 °C and above | Decomposition of carbonate |
| 900–1200 °C | Reaction of CaO with sand-aluminate compounds |
| 1250–1280 °C | Beginning of the formation of glassy phase |
| Above 1280 °C | Formação de vidro e dos compostos do cimento (clinkerization) |

more or less quickly, by the passage of a cold draft inside the clinker. Depending on the facility, the clinker presents temperature between 50 and 70 °C, on average, at the output of the cooler. Portland clinker thus obtained is conducted to the final grinding, receiving before a certain amount of gypsum, limited by the standard which aims to control in setting the start time.

## 2.2 Soil-Cement Bricks

### 2.2.1 History

The first known soil-cement application for residential building is dated to about 10,000 years in the construction of the city of Jericho, which was built entirely with soil (but the stabilizer used was animal urine and vegetable waste) (Abiko 1995). When common Portland cement is added to soil, the resulting building material is termed soil-cement and according to Neves (2000), soil-cement is a mixture of soil, cement, and water that when compressed acquires mechanical strength and durability necessary for construction purposes. Soil-cement is a very old building material and finds its roots in the changes of an even older material, soil-ash. The addition of cement to the soil results in a material that does not undergo large volume variation by the absorption and loss of humidity, does not completely deteriorate when submerged in water, and presents high compression strength and durability due to its lower permeability (Grande 2003). Soil-cement is obtained by mixing soil, pulverized and moistened at optimum moisture content, to 7–14 % Portland cement in relation to the volume of compacted soil (Vargas 1977). It is believed that British engineer H.E. Brook-Bradley, at the late nineteenth century, was the pioneer in using this mixture, initially for the treatment of road beds and tracks for horse-driven vehicles in Southern England. In Brazil, soil-cement was used in the production of road bases and studies were focused on this end. In 1948, however, the Brazilian Portland Cement Association—ABCP, suggesting another use for this material, published in its bulletin No. 54—houses were made with soil-cement walls—in which, motivated by the success achieved in some experiments, proposes to use this material for the construction of monolithic walls (Neves 1978). However, the first official record of its use in Brazil is in the building, completed in 1948, of the headquarters of the English Farm, in the city of Petrópolis —RJ (Conciani and Oliveira 2005). Soil-cement is a low-cost alternative material obtained by the mixture of soil, cement, and a little water in suitable proportions. At first, this mixture seems a wet mixture and, after compression and setting, it hardens and gains enough consistency and durability over time for many applications in rural and urban areas. Soil-cement is an evolution of past construction materials, like clay and mud. Natural adhesives of varying characteristics have been replaced by an industrial product of controlled quality: the cement. The use of soil-cement in Brazil has, since 1948, helped meeting these needs, being today already widespread. It has been demonstrated that the application of soil-stabilizing technique brings the following advantages:

- Soil-cement has been consecrated as an alternative technology for offering the main component of the mixture—soil—in abundance in nature and generally available on the construction site or close to it;
- The constructive process of the soil-cement mixture is very simple and can be conducted by unskilled labor;
- It offers good comfort conditions, comparable to brick and masonry buildings or ceramic blocks, offering no conditions for the proliferation of insects harmful to public health, meeting minimum living conditions;
- This material has good resistance and perfect waterproofing features, resisting weathering and humidity, facilitating conservation;
- The application of roughcast or plaster mortar is unnecessary due to the smooth finish of monolithic walls as a result of the perfection of pressed faces (walls) and material impermeability requiring only the application of a simple cement-based painting, further increasing its impermeability, as well as visual appearance, comfort, and hygiene;
- Low aggression to the environment, since it eliminates the firing process;
- Low transport costs when produced at the construction site;
- Low cost compared to conventional masonry.

Soil-cement has been used for decades, but its use is still very limited. As a result, entire forests are devastated to produce ceramic bricks that, after all, are more expensive. Despite these positive points, in Brazil, the interest by the soil-stabilizing method is more significant in paving works (about 90 % of the bases of our roads are made of compacted soil-cement), dams and retaining walls, with secondary application in civil construction due to the lack of technical knowledge of professionals involved in the various segments of society. Given this reality, studies aimed at developing and disseminating this technique that is needed.

## 2.2.2 Soil-Cement Strength

In the 80s, interested in spreading the soil-cement technology, company SUPERTOR manufacturer of machines and for soil-cement technology published a handbook on the operation and use of such technology. This handbook presents some factors that influence the strength properties of soil-cement brick such as:

(a) Soil characteristics;
(b) Cement content in the mixture;
(c) Degree of fineness of the cement used;
(d) Degree of homogenization of the mixture;
(e) Densification of the mixture in the pressing stage (or packing factor of the mixture);
(f) Setting time and mixture condition after pressing;
(g) Additives used.

The cement content used to stabilize the soil improves and increases the material strength and durability. The proper combination of these factors optimizes strength. It is well known that soils with higher proportion of sand in their composition, in most cases, will lead to greater soil-cement strength. The influence of other factors such as the limits of consistency, particle size distribution, and types of clay minerals should also be considered. Good homogenization of the mixture is critical. Cement should be added to dry soil and mixed until uniform color is achieved (CEBRACE 1981). Only after homogenization, water is added in adequate amounts. Resistance increases proportionally to the cement content used; however, it should be limited to an ideal content that provides the brick or blocks the required strength without unnecessary increase in the cost of the final product (ABCP 1985). Tests carried out in soil-cement specimens showed strength gains as a function of the setting time. This behavior is associated with interactions of clay mineral components and cement that, according to several authors, are little-known reactions. There seems to be some consensus that the hardening and strength gain of the mixture over time are largely associated to the reactions among clay mineral components and the lime released in the cement hydration (Segantini and Carvalho 1994). According to Ceratti and Casanova (1988), to study the strength gain of soil stabilized with cement, one must carefully study the following aspects:

• Genesis, composition and soil properties;
• Physical and chemical characteristics of soils;
• Knowledge of the cement used as binder.

Figure 2.1 shows some soil-cement bricks manufactured with different soil compositions.

**Fig. 2.1** Examples of soil-cement bricks

## 2.2.3   Criteria for the Selection of the Soil to Be Used in the Manufacture of Soil-Cement Bricks

The Brazilian Association of Technical Standards (ABNT 1989) through its NBR 10832 and 10833 standards establishes criteria for the selection of soils for use in the manufacture of solid soil-cement bricks (Table 2.2).

It is recommended the use of soils with 70 % sand and 30 % clay and 4–5 % moisture content. Sandy soils require lower amount of cement for stabilization when compared to clayey soils. CEPED suggests an assay to determine the soil suitability and its possible use for the manufacture of soil-cement bricks. It is suggested to place an amount of soil in a box of dimensions 60 × 8.5 × 3.5 cm and allow it rest for 7 days. After this period, if the contraction observed in the soil is

**Table 2.2** Criteria established by Brazilian standards for the manufacture of soil-cement bricks

| Features | Requirement (%) |
| --- | --- |
| % soil passing in ABNT 4.8 mm sieve (number 4) | 100 |
| % soil passing in ABNT 0.075 mm sieve (number 200) | 10–50 |
| Liquid limit | <45 |
| Plasticity limit | <18 |

**Table 2.3** Values established by the Brazilian standard for soil-cement bricks

| Limit values | |
| --- | --- |
| Compressive strength (MPa) | ≥2 |
| Water absorption (%) | ≤20 |

less than 2 cm in the length direction of the box and there are no cracks, the soil can be considered suitable for use in soil-cement mixtures (Neves 1989). If the soil available does not meet the proposed criteria, it can be mixed with other soils in order to obtain the necessary features. Usually, the use of soils with organic matter content less than 2 % is recommend. Regarding the cement content, the Brazilian Portland Cement Association (ABCP 2000) recommends the addition of 7–14 % of cement content to the mass, depending on the soil type. Sandy soils are stabilized with lower cement content when compared to clayey soils. Neves (1978) shows the need for wetting the bricks produced, stressing that the absence of such procedure can lead to loss of strength of about 40 %. This procedure aims to prevent water evaporation or exchange of moisture with the environment for a minimum period of 7 days. The Brazilian Association of Technical Standards (ABNT) through its NBR 10836 standard sets limit values for compressive strength and water absorption for soil-cement bricks (Table 2.3).

## 2.3  Soil-Cement Bricks Incorporated with Waste Materials

Industrial activities generate enormous amounts of solid waste, which can cause adverse effects to the environment. In Brazil, most of this residue has been improperly disposed into the environment causing environmental impacts.

Many studies have been reported in literature about soil-cement bricks with the addition of different waste materials. Table 2.4 summarizes the results obtained for some waste materials investigated in literature.

Faganello (2006) investigated the characteristics of soil-cement specimens added with gravel tailings. The results showed that the compressive strength of samples increases with increased cement contents. Lima (2010) studied the durability of green soil-cement bricks incorporated with granite waste. The authors used granite waste from industries located in the industrial district of Campina Grande-PB, Brazil. They also observed that strength decreased with increasing amounts of granite residue. When incorporated into soil-cement bricks, increased water absorption, weight loss, and volume change of bricks were observed as the waste content increased. Simone (2013) observed that the incorporation of ornamental stone waste caused changes in the technological properties of soil-cement specimens. It has been found that the ornamental stone waste tends to decline to increase the compressive strength and decrease the water absorption of test specimens with the addition of 40 % waste. All specimens with additions of 0–40 % ornamental

**Table 2.4** Comparative strength and water absorption values reported in literature for soil-cement bricks added of waste materials

| Waste materials | Compressive strength (MPa) | Water absorption (%) |
|---|---|---|
| Brita (Faganello 2006) | 1.5–3.2 | – |
| Ornamental rock (Ribeiro 2013) | 3.6–9.7 | 17–21 |
| Marble and Granite (Miranda 2007) | 2.12–3.29 | 19.52–20.46 |
| Granite (Lima 2009, 2010) | – | 16–24 |
| Eggshell (Amaral and Smith 2013) | 4–4.5 | 7.5– 21 |
| Granite (Ribeiro 2014) | 4–5.5 | 20–21 |
| Grits (Siqueira 2013) | 5–6 | 18– 22 |
| Ceramic waste (Silva et al 2014) | 0.8–3 | 16–18 |
| Rice/Brachiaria husk (Ferreira 2008) | 0.62–3 | 19.96–11.42 |

stone waste exceeded the minimum values for compressive strength set by NBR 10834 (ABNT 1994), with water absorption values below 22 %, as recommended by Brazilian standards. Matheus, 2013 studied the incorporation of eggshell waste, generated in large amounts in the food industry, into soil-cement bricks. Eggshell waste is rich in calcium carbonate ($CaCO_3$) and is considered to be a solid waste material of very complex and difficult final disposal. The results indicate that eggshell waste can be used in soil-cement bricks with excellent technical properties, in the range up to 30 wt%, as a partial replacement for Portland cement. Silva 2014 studied the incorporation of ceramic waste (construction and demolition materials) into soil-cement bricks. This type of waste presents a wide variation in chemical composition, depending on the collection of the material in the construction works, which leads to a wide variation in the properties of produced bricks. Briefly, it could be inferred that soil-cement bricks provide good conditions for the incorporation of different types of industrial waste. The addition of such wastes produces soil-cement bricks with properties within the Brazilian standards providing good opportunity for the use of these materials and consequently greater relief for landfills where these materials are usually dumped and therefore environmental improvements.

# References

Abiko AK (1995) Introduction to housing management. Sao Paulo, EPUSP (Text Technical USP Polytechnic School. Department of Construction Engineering, TT/ PCC/ 12)

Amaral MC, Smith FB (2013) Netherlands JN—Soil cement bricks incorporated with eggshell waste. Waste Mange Resour 166:137–141

Big FM (2003) Modular brick-making soil-cement by manual pressing with and without silica fume. San Carlos. Dissertation

Brazilian Association of Technical Standards—ABNT (1989) NBR 10833: massive brick manufacture and soil-cement hollow block with use of hydraulic press. Procedure. Rio de Janeiro, RJ, 3p

Brazilian Association of Technical Standards - ABNT (1994) NBR 10836, Compression resiatance and waterabsorption detremination , Rio de Janeiro,RJ, 2p

CEBRACE—Brazilian Center for School Buildings and Equipment. Solocimento in building schools—SC01. 2nd edn. Rio de Janeiro, MEC/CEBRACE 1981. 39p. il. (construction systems 2)

Cement Association of Brazilian Portland—ABCP (1985) Manufacture soil-cement bricks with the use of hand press. Sao Paulo

Cement Association of Brazilian Portland—ABCP (2000) Basic use of portland cement tab. BT-106, 7th edn. São Paulo

Cement Association of Brazilian Portland—ABCP (2002) Basic use of portland cement tab. 7th edn. São Paulo

Conciani W, Oliveira JLMD (2005) Popular Houses—options for improving the quality, in 2005

Faganello AMP (2006) Waste crushing of basaltic origin: characterization and use in soil-cement in the region of Londrina (PR). Dissertation (Master in Buildings and Sanitation Engineering) —Londrina—PR, State University of Londrina

Ferreira R, Gobo JC, Cunha AHN (2008) Rice husk Merger and pasture and their effects on physical and mechanical properties of soil-cement bricks. Eng Agric Jaboticabal 8:1–11

Grande FM (2003) Modular brick-making soil-cement by manual pressing with and without silica fume. São Carlos, Dissertation

Lima TV (2009) Soil stabilization clay for the production of ecological blocks. Engineering 34:15–26

Lima RCO (2010) Study of durability of monolithic walls and soil-cement bricks incorporated with granite waste. Thesis (MS in Civil and Environmental Engineering)—Campina Grande— PB, Federal University of Campina Grande

Miranda RAC (2007) Technical feasibility of application of marble and granite processing residue in soil-cement bricks. Dissertation (Masters in Environmental Engineering)—Goiania—GO, Federal University of Goiás

Neves CMM (1978) New materials for the improvement of public housing. CEPED—Centre for Development Research, Camacari

Neves CMM (1989) Performance walls—adopted Procedure for soil-concrete monolithic walls. CEPED—Centre for Development Research, Camacari

Neves CMM (2000) Soil cement to an environmental friendly building material. In: Symposium on construction and environment theory into practice. São Paulo

Ribeiro SV (2013) Reuse of waste rocha ornamental and soil-cement brick production. Dissertation of Magister, UENF, Brazil

Ribeiro SV (2014) Netherlands JN—Soil cement bricks incorporated with granite cutting sludge. Int J Eng Sci Innovative Technol 3:401–407

Segantini AAS; Carvalho D (1994) Behavior of small soil-cement diameter piles. X COBRAMSEF—Brazilian Congress of Soil Mechanics and Foundation Engineering. Foz do Iguaçu, vol 1, Nov 1994

Silva VM, Gois LC, Acchar W (2014) Incorporation of ceramic waste into binary and ternary soil-cement formulation for the production of solid bricks. Mater Res 17:326–331

Simone VR (2013) Reutilizacao de residuo de rochas ornamentais e producao de tijolo solo-cimento, MScDissertaton

Siqueira FB (2013) Holanda JN—Reuse of waste grits for the production of soil-cement bricks. J Env Manage 13:1–6

Vargas M (1977) Introduction to soil mechanics. McGraw-Hill, Sao Paulo

# Chapter 3
# Solid Waste Materials

**Abstract** This chapter describes two waste materials added to soil-cement bricks. Origin and characteristics of the soil drilling waste as well as the sugar cane bagasse ash.

**Keywords** Petroleum · Soil drilling waste · Sugar cane bagasse ash

## 3.1 Petroleum

Petroleum is found in nature occupying voids of porous rock called reservoir rock. Considering its organic origin, oil is a fossil fuel, probably originated from remains of aquatic animals accumulated at the bottom of primitive oceans and covered by sediment. Time and pressure of the sediment on the material deposited on the seabed turned it into homogeneous and viscous mass of black color known as oil deposits. The records of history about the use of oil date back to biblical times. Peoples of Mesopotamia, Egypt, Persia, and Judea used bitumen for paving roads, caulking of large buildings, heating and lighting homes, lubrication, and so on. The modern oil industry dates from the mid-nineteenth century. By the late nineteenth century, the United States virtually dominated alone the world oil trade largely due to the work of entrepreneur John D. Rockefeller. The American supremacy was threatened only in the last decades of the nineteenth century by the production of oil in the Caucasus fields exploited by the Noble group, with Russian and Swedish capital. In 1901, an area of a few square kilometers in the Absheron Peninsula, near the Caspian Sea, produced 11.7 million tons in the same year in the United States recorded a production of 9.5 million tons. The rest of the world produced altogether 1.7 million tons. European companies conducted extensive research throughout the Middle East and the discovery that the region had about seventy percent of world reserves caused upheaval in all levels of exploitation. Petroleum has a strong presence in the consumer society of today's world, as shown by the huge amount of processed products using inputs or directly produced from this raw material and its derivatives. The First World War highlighted the strategic importance of oil.

W. Acchar and S.K.J. Marques, *Ecological Soil-Cement Bricks from Waste Materials*, SpringerBriefs in Applied Sciences and Technology, DOI 10.1007/978-3-319-28920-5_3

The transformation of oil into war materials and the widespread use of its derivatives—it was the time when the automobile industry began to show sharp growth—caused the supply control to become a matter of national interest. The US government began to encourage the country's companies to operate abroad. Commercially, there are two types of oil: light oil (with higher proportion of gasoline) and heavy oil (with higher proportion of kerosene and fuel oils). Light oil has higher price on the world market due to the high consumption of gasoline. The world's largest oil producers are Russia, Saudi Arabia, United States, Iran, China, Canada, Mexico, United Arab Emirates, Iraq, Kuwait, Venezuela, Norway, according to the Brazilian Statistical Yearbook of Petroleum, Natural Gas, and Biofuels (ANP 2010) Brazil is in 14th place. The world has become dependent on oil and Brazil, after the 1973 crisis, started to invest in the exploitation of oil deposits in order to reduce external dependence on this very important source of raw materials. Egyptians used it to pave roads, to embalm the dead and to build pyramids, while Greeks and Romans made use of oil for military purposes. In the New World, oil was known to the pre-Columbian Indians, who used it to decorate and waterproof their ceramic pots. The Incas, the Mayans, and other ancient civilizations were also familiar with the use of oil taking advantage of it for various purposes. Oil was taken from natural exudations found on every continent. The start and support in the search process with growing affirmation of the product in modern society dates back to 1859, when commercial operation started in the United States, shortly after the famous discovery of Cel. Drake in Tittusville, Pennsylvania, an oil well only 21 m deep drilled with a steam-powered percussion system that produced 2 $m^3$/day of oil. It turned out that oil distillation resulted in products that replaced with great advantage kerosene obtained from coal and whale oil, which were widely used for lighting. These events marked the beginning of the oil era. Later, with the invention of gasoline and diesel engine, these derivatives until then despised added significant profits to the activity. By the end of the last century, oil wells multiplied and drilling with percussion method experienced its golden age. In this period, however, the rotary drilling process began to develop. In 1900 in Texas, American Anthony Lucas, using the rotary process was found oil at a depth of 354 m. This event was considered an important milestone in rotary drilling and oil history. In the following years, rotary drilling developed and progressively replaced the percussion drilling method. The improvement of projects and quality of steel, new drills, and new drilling techniques made it possible to drill oil wells with more than 10,000 m deep. Until 1945, the oil produced came from the United States, world's largest producer, followed by Venezuela, Mexico, Russia, Iran, and Iraq. With the end of World War II, a new geopolitical and economic framework was established and the oil industry was not outside the process. Still in the 1950s, the USA still held half of the world's production, but a new and potentially more vigorous oil production pole emerged in the Eastern Hemisphere. This decade also marked an intense exploration activity, and oil started being explored in the sea, with the emergence of new exploration techniques. Thus, over time, oil became the main source of energy. Today, with the advent of petrochemical industry and extensive use of oil derivatives, hundreds of novel compounds are produced, many of them are used daily such as plastics, synthetic rubbers, inks, dyes, adhesives, solvents, detergents, explosives, pharmaceuticals, cosmetics, etc. Thus, in addition to producing fuel, oil became indispensable to the facilities and amenities of modern life (Thomas 2001). The history of oil in Brazil began in 1858, when Marquis of Olinda signed Decree No. 2266 granting to José Barros Pimentel the right to extract bituminous mineral for kerosene production on land situated on the banks of the Maraú River, in the then province of Bahia. The first Brazilian well in order to find oil; however, was drilled only in 1897 by Eugenio Ferreira Camargo in the municipality of Bofete, state of São Paulo. This well reached the final depth of 488 m and, according to reports at the time, the well produced 0.5 m$^3$ of oil. Since its creation, Petrobras has discovered oil in the states of Amazonas, Pará, Maranhão, Ceará, Rio Grande do Norte, Alagoas, Sergipe, Bahia, Espirito Santo, Rio de Janeiro, Paraná, São Paulo, and Santa Catarina. Each decade in the company has been marked by very relevant facts in oil exploration in the country. The decade of 1950 was marked by the discovery of oil fields Tabuleiro dos Martins in Alagoas, and Taquipe in Bahia. In the 1960s, the oil fields of Carmópolis in Sergipe, and Miranga in Bahia. Also in Sergipe, a notable landmark of this decade was the first discovery in the sea, the Guaricema oil field.

## 3.2 Petroleum Exploration Process

The process of oil well drilling generates gravel, and percussion and rotary are the most widely drilling methods used. In the percussion method, rocks are struck by a steel-pointed drill with alternating movements causing fracturing or crushing. Periodically, it is necessary to remove the debris cut by the drill, which is achieved by lowering into the well a tube equipped with a handle at its upper end and a valve at the bottom (bucket). The bottom valve is alternately opened and closed by a protruding rod that hits against the well bottom when the bucket is being moved. This causes the entry of debris, which are removed from the well (Thomas 2001). Given its characteristics, this process is very limited, reaching maximum depths between 200 and 250 m. The rotary method uses a different technology in which the drill is rotated and pressed on the rock structures that are fragmented. These fragments are carried by a fluid (drilling fluid) that is injected inside steel pipes to the bottom returning to the surface by the annular space between the well and the outer walls of the piping. This method is used today for the drilling of oil wells (Thomas 2001). The fluid is separated from gravel by vibrating screens and returns to the tank and, if necessary, is treated, being reinjected into the well thus operating in closed system as seen. Continuous analysis of gravel allows detecting the first evidence of hydrocarbons in the rock structure.

As previously mentioned, rock fragments (gravel) are carried out by the drilling fluid to the surface of vibrating screens, which are separated from the fluid and disposed in a dike. Considering that in the activity of fixing and cleaning the probe and location, aspects with the greatest impact are: oil, gravel, and drilling fluid on the location ground and sewage treatment, which impacts pollution of soil and risk to health and environment, and regarding the gravel:

– During its removal in the process of drilling oil wells, there is pouring out of the box in which it will be disposed;
– During loading the bucket, there is overflow and leakage of collection boxes.

Since there is total removal of fluid impregnated in the gravel, they can contain contaminants such as:

(a) Heavy metals: The main risk to the environment associated with heavy metals is their water-soluble or exchangeable forms. However, these forms are present in minimum amounts in solid waste, as shown in solubility analysis according to Brazilian standard NBR 10004 (ABNT 2004), which classifies solid waste according to their potential risks to the environment and public health, so that these wastes have proper handling and disposal.
(b) High salinity: since most fluids have salt in their composition, whose objective is to minimize the swelling of perforated clay formations and promoting stability soluble salts such as sodium chloride and potassium chloride are part of the basic composition of drilling fluids, and the form of these salts in the soil dissolved in the drilling waste can bring serious consequences to the environment. Excessive concentration of soluble salt in the soil increases the osmotic potential, which is the main cause of poisoning and death of plants. The osmotic potential is the force with which dissolved constituents try to retain water molecules, i.e., salt in the soil competes with plants by water molecules. Excess salt in the soil causes plants to have premature stress by dryness, even though substantial amounts of water are available (Garcia and Cowboy 2001).

Additionally, the leaching of salt (e.g., rain) can transport salt to underground fresh water changing the water quality.

(c) Oils and greases;
(d) Elements that cause biochemical oxygen demand (BOD);
(e) Elements that cause chemical oxygen demand (COD);
(f) Elements that cause alkalinity.

During the drilling of an oil well, the waste is stored in dikes. These drilling dikes have size compatible with the final depth to be achieved in the well, and is usually between 1.0 and 1.5 $m^3$ per meter of well to be drilled. In addition to gravel, dikes also receive liquid effluents from operations (mud debris, contaminated water, cement remains). Drilling dikes must be sealed to ensure that there is no leakage of contaminants that may be deposited in them during drilling. With the completion of the drilling work, these wastes should receive adequate disposal in order to minimize damages to the environment.

## 3.3   Waste Generated in the Process of Petroleum Well Drilling

Petroleum production in Brazil grew from 750 m$^3$/day at the time Petrobras was created in 1953 to more than 182,000 m$^3$/day in late 1990s due to continuous technological advances in drilling and production (Thomas 2001). Drilling fluids are complex mixtures of solids, liquids, chemicals, and sometimes even gasses. From the chemical point of view, they can assume aspects of suspension, colloidal dispersion or emulsion depending on the physical state of components (Lima 2001). Its main purposes are: to carry fragments from drilled rocks to the surface; keep these fragments suspended in circulation stops of fluid into the well; cool and lubricate the drill; hydraulically support the well walls and contain the fluids (oil, gas or water) in the reservoir. To fulfill its goals, the fluid needs to have the capacity of not reacting with the formations it gets in contact with. Two types of formations can be found:

- Formations with active rocks are those in which rocks, due to their clayish characteristics, can interact with the fluid absorbing water and causing the hydration of clays or shales which swells the rock;
- Formations with inert rocks are those in which rocks do not suffer interaction with the water fluid such as sandstones.

The classification of a drilling fluid is a function of the main constituent of the continuous phase or dispersant as follows:

(a) Water-based fluid: water is the continuous phase and may be either fresh or salty. The primary function of water is to provide the medium dispersion for colloidal materials. These, especially clays and polymers, control viscosity among other aspects.

(b) Oil-based fluid: the continuous phase is oil which may contain up to 45 % of water (inverse emulsion in which water droplets are encapsulated by oil having a greater difficulty to interact with active rocks). The main characteristics of oil-based fluids, which gives them advantages over water-based fluids are: high degree of inhibition in relation to active rocks; very low corrosion rate, controllable properties above 175 °C; high degree of lubricity; wide density variation range (from 0.89 to 2.4 kg/l); very low solubility of inorganic salts. However, oil-based fluids have some drawbacks and the most significant is the increased pollution levels and greater initial cost. Recently, much progress has been made regarding the search for new oil-based systems such as mineral and synthetic oils, less polluting than diesel, and with much greater degree of biodegradability.

## 3.4  Soil Drilling Waste from Petroleum Industry (SDW)

Soil drilling waste from petroleum industry (SDW) are mixtures of small rock fragments impregnated with drilling fluid used to lubricate and cool the drill during drilling (Leonard and Stegemann 2010). When improperly disposed, SDW pollute soil damage the landscape and constitute an environmental problem if disposed in landfills without prior treatment. The amount of oil drilling waste generates significant negative impacts (Page et al. 2003). Theoretically, the gravel volume generated during the drilling of an oil well is the geometric volume of the perforated drum (called nominal well volume). But the calculation of the gravel volume produced uses a safety factor of about 20 % due to possible landslides of rock formations into the well during drilling. The average waste volume generated in onshore drilling is 13 m$^3$ for each 100 m of drilling advance depending on the well diameter (Petrobras 2010). Drill gravels tend to have angular shape and particle size distribution ranging from gravel to clay. The physical composition of gravels reflect the geologic materials that make up the subsoil being drilled, as well as specific solid components and other chemicals that originally compose the drilling fluid (Charles and Sayle 2010). The presence and concentration of gravel contaminants depend on the fluid used, the geological formation being drilled, the well phase, and the water used in the preparation of fluids. The main contaminating agents can be divided into hydrocarbons, water-soluble salts and in some cases heavy metals (Reis 1996). Hydrocarbons may contaminate gravels during their discovery as crude oil or if contained in fluids. They are partially removed by the solids control system, though such removal is not complete. Crude oil contains thousands of different types of hydrocarbon molecules. The toxicity and potential environmental impacts vary considerably for different molecules. A number of bioassays have been conducted to determine the toxicity of hydrocarbons to marine animals. The hydrocarbon toxicity that has been found is quite variable and generalizations cannot be easily made. Factors affecting toxicity include: molecular weight, hydrocarbon family, exposed organism and its stage in the life cycle. For mixtures of hydrocarbons such as crude oil, toxicity is also dependent on exposure history. For hydrocarbons of same family, toxicity tends to increase as molecular weight reduces. Small particles tend to be more toxic than larger ones. The most toxic hydrocarbons are those having increased water solubility. High solubility makes the molecule more accessible for absorption by plants and animals (Reis 1996). Soluble salts such as potassium and sodium chloride are part of the basic composition of drilling fluids used in oil and gas wells, and the disposal of these salts in the soil dissolved in gravels can bring serious consequences to the environment. Excessive concentration of soluble salt in the soil increases the osmotic potential, which is the main cause of damage and death of plants. The osmotic potential is the force with which dissolved constituents try to retain water molecules, i.e., salt in the soil competes with plants by water molecules. Excess salt in the soil causes plants to have premature stress by dryness, even though substantial amounts of water are available. Additionally, the leaching of salt (e.g., rain) can transport salt to underground fresh water, changing the water quality (Reis

1996). Since there is no total removal of the fluid impregnated in gravels, they may have contaminants. These contaminants depend on the chemical composition of the drilling fluid and composition of the rock formation. Some studies (Abbe et al. 2009, Pires 2009, Medeiros 2010 and Leonard 2010) described the chemical composition of drilling gravels expressed in the most stable oxides, and heavy metals detected in SDW are summarized in Table 3.1.

The most common measure of environmental impact potential of a material is its toxicity. Toxicity is defined as the potential property that the toxic agent has to cause, to a greater or lesser extent, an adverse effect as a result of its interaction with a given organism. Toxicity occurs when some material causes a deleterious effect on an organism, population, or community. The toxicity of a substance is a measure of how it harms the life and health of living beings following exposure to the substance. Two types of toxicity measurements are commonly used: dose and concentration. Dose is the concentration of a substance that has been absorbed into the tissue of the species under study, while concentration is a measure of the concentration of a substance in the environment of living species and also includes an exposure time interval. Toxicity bioassays are conducted in animals and the results are extrapolated to create guides containing the maximum allowable values for each element or substance (Reis 1996). In Brazil, NBR 10004 (ABNT 2004) contains the tables with the maximum limits in the extracts obtained in leaching and solubilization tests, which are estimated for the exposure level that probably will not result in harmful effects throughout lifetime. However, these limits are based on water drinkability standards. Susich and Schwenne (2004) discuss these limits because they disregard the solubility of these metals. There are metals such as galena (lead sulfide) containing about 87 % lead; however, galena is essentially insoluble resulting in little risk to those exposed to it. They also propose an increase in the maximum limits allowed by about 10 times the current limits, which would be much lower than the limits considered toxicity of metals. A standard based on the total concentration of

**Table 3.1** Heavy metals detected in SDW

| Determinações (%) | Abbe et al. (2009) | Pires (2009) | Medeiros (2010) | Leonard and Stegemann (2010) |
|---|---|---|---|---|
| $SiO_2$ | 47.60 | 43.96 | 36.5 | 60.4 |
| $Al_2O_3$ | 13.54 | 21.48 | 11.5 | 10.4 |
| $Fe_2O_3$ | 6.34 | 5.40 | 4.5 | 4.9 |
| BaO | 11.39 | 2.38 | N.A. | N.A. |
| CaO | 2.78 | 18.12 | 35.3 | 2.5 |
| MnO | 0.17 | N.A. | 0.09 | 0.06 |
| MgO | 2.31 | N.A. | N.A. | 2.0 |
| $K_2O$ | 2.33 | 4.51 | 2.7 | 1.7 |
| $Na_2O$ | 1.17 | N.A. | N.A. | 2.4 |
| $TiO_2$ | 0.65 | N.A. | 0.81 | 0.6 |
| $P_2O_5$ | 0.10 | N.A. | N.A. | 0.1 |

Fonte: Fialho (2012)

metals in rocks is very conservative, or rather unreal. The main risk to the environment associated with heavy metals is related to its water-soluble or exchangeable form (Pires 2009). Humans require only small amounts of some of these metals and excesses may have adverse effects on human health. Another criticism by Susich and Schwenne (2004) is about not considering the physical characterization of gravel to determine the availability of metals, as fractions smaller than 50 mesh (297 mm) are considered of higher risk because small particles have larger surface area per mass unit and present high potential to leach metals.

## 3.5   Sugarcane Bagasse Ash

### 3.5.1   Sugarcane Bagasse

Sugarcane is an agricultural product that originated from Southeastern Asia and grown in Brazil since Portuguese colonization. Sugarcane was brought to Brazil in 1532 by Martin Afonso de Sousa, and the region in northeastern Brazil called "Zona da Mata" was the main production region, and since then sugarcane started to have significant importance for the country. The importance of sugarcane is due to its diverse use: as raw material for the production of alcohol, sugar, Brazilian spirit. However, the main destination of sugarcane in Brazil is the manufacture of sugar and alcohol. However, there is Brazilian and international technology through which various by-products such as bagasse, straw, honey, filter cake, and vinasse are used (Ripoli 2004). It is noteworthy that, in sugar mills alcohol is manufactured with honey resulting from sugar production; thus the majority of plants have distillery to manufacture alcohol (Pinto 1999). The use of sugarcane to produce ethanol or sugar has varied over time. In the 1970s, about 90 % of sugarcane was processed into sugar; but position was reversed in the 1980s, after the creation of PROÁLCOOL program, when about 80 % of sugarcane was used to produce alcohol (Brazil 2007). Thus, the number of independent distilleries that produce alcohol directly from sugarcane juice doubled without worrying about sugar production (Pinto 1999). The participation of sugarcane in the energy sector takes into account not only the alcohol consumed by motor vehicles, but also the use of bagasse at mills. Bagasse is the solid waste generated during the extraction of juice from the crushing of sugarcane to produce sugar and alcohol. Its chemical composition varies with species of sugarcane being cultivated, types of herbicides and fertilizers used, and natural factors such as climate, soil, and water (Cordeiro 2006).

The main solid waste from the sugar and alcohol industry are sugarcane straw, sugarcane bagasse, and sugarcane bagasse ash being classified according to nature as class II; and packaging of pesticides are classified as Class I according to chemical components (FIESP/CIESP 2001). Sugarcane bagasse can be turned into hydrolyzed bagasse for animal feed or can be burned in the boiler, along with straw in the case of electricity cogeneration (FIESP/CIESP 2001; Pinto 1999; Ripoli 2004).

## 3.5.2   Use of Sugarcane Bagasse

During sugar and ethanol production, sugarcane bagasse is generated as a by-product resulting from the extraction of juice from sugarcane in the milling process. The amount of bagasse extracted reaches approximately 30 % of crushed sugarcane and has been used as an energy source, since about 95 % of this biomass is burned in boilers to generate steam to produce sugar and alcohol (Paula et al. 2009). The main by-products of sugar and alcohol production include: vinasse (the remaining juice after distillation is fractionated in ethanol manufacturing), filter cake (consists of a solid material that is trapped in the filter after sugar fermentation during filtering of the liquid that still contains sucrose), and bagasse. The latter is a solid residue composed of lignin and fresh cellulose, consisting of 45 % lignocellulosic fibers, 50 % moisture, 2–3 % insoluble solids, and 2–3 % water-soluble solids. Chemically, it is composed of cellulose, hemicellulose and lignin, with 41, 25 and 20 %, respectively, based on the dry weight of bagasse (Zardo 2004). In recent times, research to optimize processes in order to enable the use of these by-products and the development of other technologies to increase their value has been the subject of attention of sugar and alcohol industries. Cordeiro (2006) states that for every ton of sugarcane, about 260 kg of bagasse with 50 % humidity are produced. Among wastes produced in a sugarcane plant, bagasse is one of the most attractive due to its calorific value, making it the main energy source in the production process of sugar and alcohol manufacturing (Cordeiro et al. 2008; Souza et al. 2007). Although it is also considered as one of the major wastes of the national agro-industry, its industrial application ranges from the manufacturing of composite for animal feed, fertilizer, and biogas to use as raw material for plywood and chemical industry in general. In early twenty-first century, its use was directed to energy production (thermal and mechanical), known as cogeneration (Souza and Azevedo 2006). By definition, cogeneration is the simultaneous generation of thermal and mechanical energy from a single fuel (natural gas, wood waste, rice husk, sugarcane bagasse, straw, pointers, etc.)

## 3.5.3   Sugarcane Bagasse Ash

Among wastes, mineral ashes from different agroindustrial activities stand out, which have high percentages of silica and other oxides and may be used as pozzolans. The pozzolan property is the capacity to react with calcium hydroxide released during the cement hydration process forming stable compounds of binder power such as silicates and calcium aluminate hydrates (Oliveira et al. 2004). During extraction of sugarcane juice, large amounts of bagasse are generated (approximately 30 % of crushed sugarcane), which is the biomass of paramount importance as an energy source. About 95 % of all sugarcane bagasse produced in Brazil are burned in boilers to generate steam, producing ash as waste, which disposal does not follow in most cases, favorable practices and can be source of serious environmental problems. Sugarcane bagasse ash (SBA) consists primarily of silica, $SiO_2$, and has potential to be used as a mineral addition replacing part of the cement in mortars and concretes (Cordeiro et al. 2008; Paula et al. 2009). Sugarcane bagasse is burnt in boilers by means of a system called cogeneration, which consists of the production of steam to be supplied to mechanically driven turbines such as pumps, mills, defibrillators, and also electrical power generators; and the demand of the processing of sugarcane juice is supplied with the steam that comes out of the turbine called exhaust steam (Fiomari 2004). The electricity produced can meet all or part of the plant needs enabling its self-sufficiency and also generating exportable surpluses to power distribution companies. Cogeneration acquired importance in the 1980s, although the systems of that time were considered inefficient since the turbines used generated electricity only for own consumption, and the working parameters of steam generated by boilers were 2156 kPa of pressure and 290 °C of temperature (Fiomari 2004). With the possibility of exporting electricity, in addition to market competitiveness, mills began to worry about the efficiency of their thermal machines making it necessary the modernization of their industrial park. To meet the technical requirements of such machines, the levels of pressure and temperature of the steam generated by the boiler had to be changed to 4214 kPa and 420 °C, respectively (Fiomari 2004). Cordeiro (2006) reports, higher the temperature inside the boiler and/or heat exposure time, the greater the amount of carbon released producing ash with different colors. The calcination of bagasse in the boilers results in heavy ash and fly ash, which although not released directly into the air, can pollute the environment if improperly disposed after cleaning the plant boiler (Borlini et al. 2006). For waste generation estimate, FIESP/CIESP (2001) adopted the following proportions: each ton of sugarcane produced generates 260 kg of dry bagasse and each ton of bagasse that feeds plant boilers, considering its use percentage at 95 %, generates 23.8 kg of sugarcane bagasse ash. Along with filter cake and straw, sugarcane bagasse ash is forwarded to be used as fertilizer in sugarcane crops, despite being poor in nutrients.

### 3.5.4   Chemical Composition of Sugarcane Bagasse Ash

For Cordeiro (2006), ash has as main chemical silica ($SiO_2$), usually in amounts above 60 % by weight. The author also points out that controlling burning conditions allows keeping silica contained in the bagasse in the amorphous state, the main characteristic that enables using waste as pozzolan and consequently reducing costs and environmental impact related to its disposal in the environment. Furthermore, ash can add economic value to the residue and provide technical and environmental advantages both in the addition as in the partial replacement of Portland cement (Cordeiro 2006). It is noteworthy that the high silicon content present in SBA is absorbed from the soil by the roots of sugarcane plants as monosilicic acid ($H_4SiO_4$) and subsequently transpiration (water coming out from the plant) is held on the outer wall of the epidermis cells in the form of silica gel (Borja 2011). The accumulation of silica between cuticle and the wall of epidermal cells serves as a physical barrier to the penetration of pathogenic fungi and reduces loss of water by transpiration (Barboza Filho 2002). According to Borja (2011), another possible source of silica is sand coming from sugarcane crop that is not completely removed during the washing step. Typically, these ashes are used as fertilizer in sugarcane crops, in parallel with the filter cakes and straw (Cordeiro 2006). However, due to the lack of mineral nutrients for their use as fertilizer, often these ashes are discarded in the environment without proper disposal, even though it is of difficult degradation (FIESP/CIESP 2001). With the increase in the sugar-cane industry in Brazil, an increase in the occurrence of problems related to the disposal of waste produced by the sector was also observed. Therefore, many researchers have sought to develop alternatives for sustainable destination of the material and also a way to add value to this sugarcane industry by-product. The civil construction industry is one of the strongest candidates for the destination of this waste by the incorporation of these into cementitious matrixes. In addition to being the sector that most consumes natural raw materials, it is also responsible for the consumption of 4.5 % of total world energy. However, most of this energy, about 84 %, is used in the manufacture of materials. For ash to be used as a mineral addition, it must be derived from industry or vegetable with high contents of silicates produced in the amorphous state and with appropriate particle size (John et al. 2003). In the use of ash as pozzolanic addition, it becomes necessary to use some specific grinding and burning procedures. Importantly, such procedures can lead to economic infeasibility due to the low prevailing reactivity in most cases. For the silica content to become amorphous phase, ash must be fired at temperatures below 700 °C for one hour (Macedo 2009). However, Martinera Hernandez et al. 2000 state that the combustion temperature of agricultural waste should be between 400 and 800 °C in order to avoid the formation of crystalline phases of silica (product from the high combustion temperatures), and below 600 °C, the per-centage of amorphous silica is high. Sugarcanne bagasse ash with firing tempera-tures above 800 °C shows high crystallinity with quartz and mullite peaks. Cordeiro et al. (2009) identified that the temperature of 600 °C is the most suitable to produce predominantly pozzolanic sugarcane bagasse ash. Although there is no consensus on the temperature and the ideal firing time of bagasse to obtain an amorphous material, studies have shown the importance of controlling the firing of ashes to optimize their reactivity (Martinera Hernández et al. 2000; Nehdi et al. 2003). Sugarcane bagasse ash has high silicon content, usually above 60 % (by mass) and thus presents pozzolanic activity (Martinera Hernández et al. 2000; Cordeiro et al. 2008; Cordeiro 2006). Silicon is absorbed from the soil by sugarcane plant roots as monosilicic acid ($H_4SiO_4$), and after water loss by transpiration it is deposited on the outer wall of epidermal cells as silica gel (Cordeiro et al. 2009). Ashes are distinguished by their mineralogical characteristics, that is, the form in which silica is found, amorphous or crystalline.

# References

Abbe OE et al (2009) Novel sintered glass-ceramics from vitrified oil well drill cuttings. J Mater Sci 44:4296–4302

ANP—National Agency of Petroleum (2010) Statistical yearbook Brazilian petroleum. Natural Gas and Biofuels, Rio de Janeiro

Barboza Filho MP, Prabhu AS (2002) Calcium silicate application on rice—Technical Circular 51. EMBRAPA, Santo Antônio de Goiás 4p

Borja EV (2011) Clay adding the effect expanded and mineral additions in the design of lightweight structural concrete. Doctoral Thesis 2011. Federal University of Rio Grande do Norte

Borlini MC, Mendonça JLCC, Vieira CMF, Monteiro SN (2006) Influence of sintering temperature on the physical, mechanical and macro-structural red ceramic incorporated with ash residue of sugarcane. Matter (Rio J.) 11(4):433–441p

Brazilian Association of Technical Standards - ABNT NBR 10004, Rio de Janeiro-RJ, 2004

Brazil. Law No. 12.305 of 2 August 2010 establishing the National Solid Waste Policy. Official journal [of] the Federative Republic of Brazil, Brasilia, 3 Aug 2007

Charles M, Sayle S (2010) Offshore drill cuttings treatment technology evaluation. SPE publication 126333. SPE international conference on health, safety and environmental in oil and gas exploration and production held in Rio de Janeiro, Brazil, 12–14 April 2010

Cordeiro GC (2006) Use of ultrafine ash from bagasse from sugarcane and rice husk as mineral additives. Thesis (Doctorate in Civil Engineering)—Rio de Janeiro—RJ, Federal University of Rio de Janeiro, UFRJ, 445p

Cordeiro GC, Toledo Filho RD, Tavares LM, Fairbairn EMR (2008) Pozzolanic activity and filler effect of sugar cane bagasse ash in Portland cement and limemortars. Cem Concr Composites, Barking 30:410–418

Cordeiro GC, Son RDT, Fairbairn EMR (2009) Bagasse gray characterization of cane sugar for Employment as Pozzolan in cementitious materials. New Chem J 32(1)

Fialho PF (2012) Gravel drilling oil and gas wells. Application potential of the study in concrete—Dissertation (Master in Civil Engineering)—Federal University of Espirito Santo, Technological Center. Espirito Santo

FIESP/CIESP (2001) Energy supply expansion from biomass (bagasse from sugarcane). FIESP/CIESP, Sao Paulo, p 85p

Fiomari MC (2004) Energy analysis and exergy of a sugar-ethanol plant in the oeste paulista system with co-generation of energy in expansion. 130f. Dissertation (Masters in Mechanical Engineering)—Faculty of Engineering, Universidade Estadual Paulista, Ilha Solteira

Garcia RL, Cowboy RLC (2001) Feasibility of the implementation of land treatment technologies, dilution and burial roadspreading for disposal/drill cuttings remediation in UN-BA. Leopoldo Research and Development Center A. Miguez de Mello, Rio de Janeiro. Technical Communication CT BIO 88/2001

John VM, Cincotto MA, Silva MG (2003) Gray and alternative binders. In: Freire WJ, Beraldo AL (ed) Technology and alternative building materials. UNICAMP, Chap. 6, Campinas, pp 145–190

Leonard SA, Stegemann JA (2010) Stabilization/solidification of petroleum drill cuttings. J Hazard Mater 174:463–472

Lima HRP (2001) Drilling basics. Handout of the training course of petroleum engineers Petrobras, North-Northeast Human Resources Development Center, Salvador 2001

Macedo PC (2009) Evaluation of the performance of mortars with addition of gray sugarcane bagasse. Dissertation (Master). São Paulo State University, Ilha Solteira, SP. p 83

Martinera Hernandez JL, Betancourt Rodriguez S, Middendorf B, Rubio A, Martinez Fernandez L, Lopez Machado I, Gonzalez Lopez R (2000) Pozzolanic properties of the waste (primeira part) sugar industry. Build Mater 50(260):71–78

Medeiros LC (2010) Gravel added drilling the Potiguar Basin in clays for use in ceramic materials: influence of concentration and firing temperature. 2010. 91f. Dissertation (Masters in science and engineering materials). Graduate programs in science and materials engineering, Federal University of Rio Grande do Norte. Natal

Nehdi M, Duquette J, Damatty EL (2003) Performance of rice husk ash produced using the new technology the mineral admixture in concrete. Science Direct, Pergamon. Cem Concr Res 33:1203–1210

Oliveira M P, Nobrega AF, Campo MS, Barbosa NP (2004) Study of calcined kaolin as a partial replacement of Portland cement material. In: Proceedings of Brazilian materials conference and non-conventional technologies: housing and social interest infrastructure Brazil—NOCMAT 2004 Pirassununga. USP, Pirassununga

Page PW, Greaves Chris, Lawson R, Hayes S, Boyle F (2003) Options for the recycling of drill cuttings. SPE Publication 80583. SPE/EPA/DOE Exploration and Production Environmental conference held in San Antonio, Texas, USA, 10–12 Mar 2003

Paula MO, Tinoco IFF, Rodrigues CS, Silva ENS, Souza CF (2009). Potential of gray bagasse sugarcane as Portland cement partial replacement material. Available at: <http://www.scielo.br/pdf/rbeaa/v13n3/v13n03a19.pdf>

Petrobras (2010) Exploration and production of oil and gas production. Available at: http://www.petrobras.com.br/pt/quem-somos/perfil/atividades/exploracaoproducao-petroleo-gas, 2010

Pinto CP(1999) Anaerobic digestion technology of vinasse and development sustentável. 1999. 147f. Dissertation (Master in Energy Systems Planning)—State University of Campinas, Campinas

Pires JPM (2009) Use of oil-well drill cuttings to produce red ceramics. 2009. 173f. Tese (Doctorate in Civil Engineering)—Pontifical Catholic University of Rio de Janeiro. Rio de Janeiro

Reis J (1996) Environmental control in petroleum engineering, 1st edn. Gulf Publishing, Houston, Texas

Ripoli TCC, Ripoli MLC (2004) Sugarcane biomass: harvesting, energy and environment. T. C. C. Ripoli, Piracicaba, p 302p

Souza GN and Azevedo PF (2007) Denvolvimento mortars with partial replacement of Portland cement by Ash Waste from bagasse cane sugar. In: Proceedings of international congress of concrete, 49, Bento Gonçalves. IBRACON, Sao Paulo

Souza ZJ, Azevedo PF (2006) Generation of electric energy surplus in the sugar-alcohol sector: a study from São Paulo plants. Rev Econ Sociol Rural, Brasília, 44(2)

Susich ML, Schwenne (2004). Onshore drilling waste management: beneficial reuse of cuttings. SPE publication 86732. spe international conference on health, safety, and environment in oil and gas exploration and production, 29–31 Mar 2004, Calgary, Alberta, Canada

Thomas JE (2001) Fundamentals of petroleum engineering. Publisher Interscience, Rio de Janeiro

Zardo AM (2004) Using the gray sugarcane bagasse as "filler" in composite fiber cement. In: Proceedings of I Latin American meeting on sustainable construction. Sao Paulo

# Chapter 4
# Using Oil Drilling Waste in Soil-Cement Formulations

**Abstract** This chapter shows the preparation of soil-cement bricks and physical and mechanical properties of soil-cement bricks incorporated with oil drilling waste from the petroleum industry.

**Keywords** Soil-cement bricks · Physical properties · Mechanical properties

## 4.1 Raw Materials

For this work, soil samples from the municipality in São José do Mipibu/RN were collected at the banks of the Baldun River, as can be seen in Fig. 4.1. About 200 kg of soil was collected and taken to the Laboratory of Physical Properties of Ceramic Materials—UFRN, which was properly characterized.

Then, a test called Box Assay, suggested by Centre for Development Research, was performed to evaluate the expansion, contraction, and plasticity characteristics, and for that, an amount of dried and sieved soil mixed with 20 % by weight of water was placed in a box divided into five compartments with the following dimensions 60 × 8.5 × 3.5 (Fig. 4.2), leaving in the shade for a period of 7 days. The scale used to dose the soil and water was Precision PR 1000, with maximum weight of 1000 g and accuracy of two decimal places.

After 7 days from the curing period, its retraction was measured and considering the soil characterization and the results of the test described above, it became clear that it was not necessary to correct the soil for use, as it showed contraction rates less than 2.0 cm, within standards established by legislation (Fig. 4.3).

The soil drilling waste sample was provided by a oil well company from the state of Rio Grande do Norte.

© The Author(s) 2016
W. Acchar and S.K.J. Marques, *Ecological Soil-Cement Bricks
from Waste Materials*, SpringerBriefs in Applied Sciences
and Technology, DOI 10.1007/978-3-319-28920-5_4

**Fig. 4.1** Soil extraction site

**Fig. 4.2** Box used for the soil contraction assay

## 4.2    Preparation of Specimens

First, raw materials collected were crushed in a ball mill in order to reduce the size of coarse particles, being then sieved on a 4.8 mm mesh sieve (Figs. 4.4 and 4.5).

The constituents of specimens, soil, cement, and soil drilling waste were separated after weighing for subsequent conformation. The mixture of mortar components was performed manually by placing the cement in a container in order to manually crush small clods present in the material. Subsequently, soil was added (in the formulations that had this constituent), and then the waste with manual

**Fig. 4.3** Soil after contraction test showing cracks

**Fig. 4.4** Dry soil after being sieved on a 4.8 mm mesh sieve

homogenization so that the final product had uniform color. After this step, drinking water was gradually added and new homogenization was performed. The mixture was transferred from the homogenization container to the conformation matrix; the press mold that gives shape to the element that after pressing is expelled, as can be seen in Fig. 4.6.

Upon completion of the conformation of specimens, elements were placed over a flat surface; after 6 (06) hours from molding, and during the first 7 (07) days, the elements were kept moist by successive wettings (every 2 h) with spray to ensure adequate curing. Figure 4.7 shows the soil-cement-waste specimens after pressing.

**Fig. 4.5**  Soil drilling waste after being sieved on a 4.8 mm mesh sieve

**Fig. 4.6**  Conformation of specimens

**Fig. 4.7**  Soil-cement-waste brick after conformation

**Table 4.1** Formulations used in the experiment

| Formulation |
| --- |
| Soil + 14 % cement |
| Drilling waste + 14 % cement |
| Soil + 14 % cement + 80 % waste |
| Soil + 14 % cement + 70 % waste |
| Soil + 14 % cement + 60 % waste |
| Soil + 14 % cement + 50 % waste |

Table 4.1 shows the formulations investigated in the study. It was observed that a constant amount of cement of 14 % and variable amounts of waste were used. For comparison purposes, a mixture using only gravel and cement without the addition of clayey soil was prepared.

## 4.2.1   Methodology and Experimental Procedure

Figure 4.8 illustrates the experimental procedure, showing all steps carried out in the experiment.

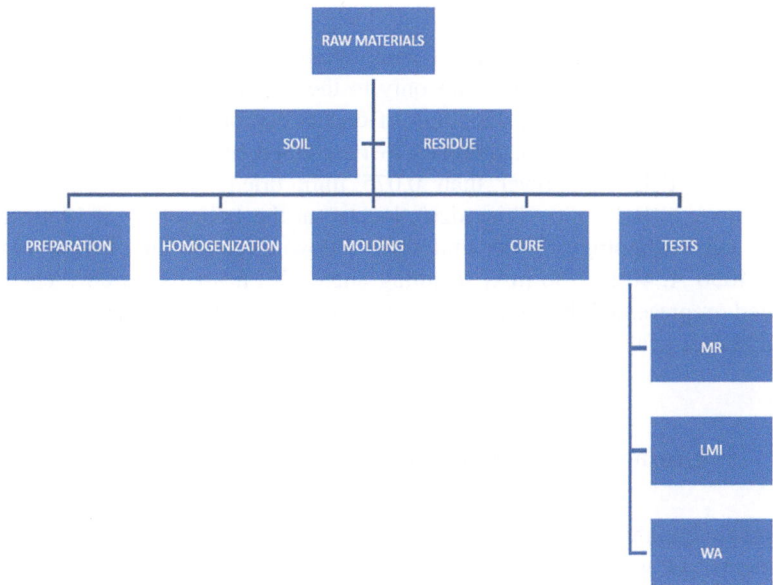

**Fig. 4.8** Experimental procedure

## 4.3   Characterization of Raw Materials

The raw materials used were characterized by the following techniques: particle size analysis (PS); mineralogical analysis by X-ray diffraction (XRD); chemical analysis by X-ray fluorescence (XRF); scanning electron microscopy (SEM), thermogravimetric analysis (TGA) and differential thermal analysis (DTA).

### 4.3.1   Particle Size Analysis

The particle size distribution of soil drilling waste and sugarcane bagasse ash samples, classified by sieving # 35, was held at the Laboratory of Materials Technology, Department of Chemical Engineering, Federal University of Rio Grande do Norte, using a laser granulometer based on light diffraction and scattering properties, model 1180L, manufactured by company Cilas, with range from 0.04 to 2500 μM, and the Particle Expert software was also used, which calculates the statistical distribution of a population of particles using complex mathematical transformations (inverse Fourier transform), using water as dispersion medium.

### 4.3.2   X-ray Diffraction (XRD)

X-ray diffraction (XRD) is the most comprehensive method for determining the mineralogy of raw materials due not only to the possibility of identifying mineral species but also for allowing the study of crystallographic characteristics of these minerals (Barba et al. 1997). Samples to be studied by XRD were classified by sieving in particle size lower than 0.075 mm, dried, and analyzed on X-ray diffractometer (XRD-6000, Shimadzu), located at the Laboratory of Metrology of the Gas and Technology Center under the following conditions: Cu-Ka radiation ($\lambda = 1.54056$ Å, 40 kV, 30 mA) scanning angle ($2\theta$) from $10°$ to $80°$. The results generated, represented by mineralogical phases of samples, were analyzed by comparison of peaks generated with the standard charts of the ICSD database.

### 4.3.3   Scanning Electron Microscopy (SEM)

The knowledge about the microstructure and individual properties of the mortar constituents, as well as the relationship between them, helps controlling their properties (Mehta and Monteiro 1994). Samples studied by scanning electron microscopy (SEM) were analyzed in a scanning electron microscope model MEV SSX 550 superscan—Shimadzu, at the Laboratory of Chemistry of the

Federal Institute of Alagoas in order to verify the behavior of wastes and if the incorporation of gravel and SBA caused significant changes in the mass. Due to the low conductivity of compounds, it was necessary to ground fragments with metal strip between the surface of the sample and the sample holder followed by the metallizing process. Images were acquired using the secondary electron detector (SE), and microanalyses of some samples in specific regions were obtained using energy dispersive X-ray spectroscopy (EDS).

### 4.3.4   Thermal Analysis

Thermal analysis is widely used in the complementation of the characterization of raw materials. The main thermoanalytical techniques used in materials are: thermogravimetric analysis (TGA) and differential thermal analysis (DTA). DTA detects energy changes occurring in the material during heating or cooling, resulting from four main causes: phase transitions, decompositions in the solid state, reactions with an active gas such as oxygen, and 2nd order transitions (entropy change without enthalpy change). In TGA, the thermogravimetric curve shows mass changes that occur during the heating of a material and may have two causes: decomposition or oxidation. TGA provides more limited information compared to DTA; however, for the quantitative analysis of certain substances, the information obtained can be more accurate. Thermogravimetric curves were obtained using a thermogravimetric analyzer (TGA-50, Shimadzu) in air atmosphere at a heating rate of 10 °C/min.

### 4.3.5   X-ray Fluorescence

Determining the chemical composition of materials is of great importance for their characterization. This analysis is of little use if some physical data are not provided or without knowing the mineralogical species present in raw materials. For being a very reproducible, fast, and accurate instrumental technique, X-ray fluorescence (XRF) is the method of choice for determining the chemical composition of materials (Barba et al. 1997). In this work, before being analyzed, samples were classified by sieving in particle size lower than 0.074 mm, corresponding to ABNT # 200 sieve and dried in an oven (110 °C) for 24 h. Tests used X-ray fluorescence spectrometer (EDX-700, Shimadzu) under vacuum atmosphere using the semi-quantitative method for the determination of elements present in samples. The results in the form of oxides are limited to the identification of chemical elements from the periodic table.

### 4.3.6   Determination of the Amounts of Cement and Water

Regarding the cement content, the Brazilian Association of Portland Cement—
ABCP recommends using 7 % as the lowest cement content to be adopted and
Vargas (1977) recommends "from 7 to 14 % of Portland cement in relation to the
volume of compacted soil". After conducting trials in over sixty soil samples from
the metropolitan area of Salvador evaluating the strength and durability as a
function of the use of various cement contents, CEPED concluded that to obtain
compressive strength equal to or greater than 2.0 MPa the cement content should be
14 %. The cement used for the production of solid and hollow brick was CP II-Z
with the addition of 6–14 % of pozzolan material, making the cement less per-
meable, ideal for underground works, especially in the presence of water.
Cement CP II-Z may also contain carbonate materials (fillers) at a maximum limit
of 10 % by mass. The choice of CP II-Z was related to the addition of pozzolanic
material, since the cement was partially replaced by SCA. According to Marques
(2010), the amount of water to be added was defined based on the Proctor Normal
compaction test performed on the soil-cement mixture, where the optimum moisture content was the parameter that defines such dose. Calmon et al. (1998) and
Ferraz (2001) report that the use of the Proctor Normal compaction test NBR 7182
(ABNT 1986) enables the determination of the amount of water to be added in the
mixture (optimum moisture content) in order to obtain maximum specific mass. The
effect of the molding moisture content on the compression strength of soil-cement
mixtures, the compressive strength reaches a maximum value and decreases similar
to the compression curve, i.e., maximum strength occurs at optimal moisture levels
(Lopes 2002). This parameter is closely related with soil particle size and plasticity
(Ferraz et al. 2005) and, as the cement content to be added to the mixture (Ferraz
et al. 2001), it grows with the amount of clay.

## 4.4   Physical Properties

### 4.4.1   Liquidity Limit and Plasticity Limit

Two parameters can be used as a criterion for the selection of appropriate soils for
use in soil-cement bricks: particle size limitation and plasticity. Thus, based on the
results of tests, CEPED recommends specification for the liquidity limit lower than
45 %. Similarly, the Brazilian Association of Technical Standards—ABNT,
through its NBR 10833 standard, establishes criteria for the selection of soil to be
used in the manufacture of massive soil-cement bricks, where the liquidity limit
must be lower than 45 % and the plasticity limit lower than 18 %. Table 4.2 shows
the liquidity and plasticity limits for the soil used and the plasticity index showed,

**Table 4.2** Liquidity limit and plasticity limit

|  | (%) |
|---|---|
| Liquidity limit | 28.4 |
| Plasticity limit | 17.8 |
| Plasticity index | 10.6 |

in compliance with NBR 10833 standard that the soil is suitable for use in the manufacture of soil-cement bricks, with no need for correction to achieve better results.

## 4.4.2 Chemical Analysis by X-ray Fluorescence (XRF)

Table 4.3 shows that the chemical characterization of soil is essentially constituted by $SiO_2$, $Al_2O_3$, and $Fe_2O_3$ oxides.

Table 4.4 shows the chemical composition of the soil drilling waste. In addition to $SiO_2$, $Al_2O_3$, and $Fe_2O_3$ oxides, there are high contents of CaO from carbonaceous rocks of petroleum reservoirs.

## 4.4.3 X-ray Diffraction (XRD)

Figure 4.9 shows the XRD pattern of soil. It was found that the sample consists of the following crystalline phases: kaolinite and quartz. The predominant clay mineral is kaolinite ($Al_2(Si_2O_5) \cdot (OH)_4$) as evidenced by its quite intense characteristic peaks and well defined form, as well as for the characteristic quartz peaks ($SiO_2$).

Figure 4.10 shows the X-ray diffraction of the soil drilling waste, where characteristic peaks of quartz ($SiO_2$), calcite ($CaCO_3$), dolomite ($CaMg(CO_3)_2$), and microcline ($KAlSi_3O_8$) were identified.

**Table 4.3** Chemical analysis of soil by fluorescence

| Soil | | | | | | | | | | | | |
|---|---|---|---|---|---|---|---|---|---|---|---|---|
| Oxides | $SiO_2$ | $Al_2O_3$ | $Fe_2O_3$ | $TiO_2$ | MgO | $K_2O$ | $ZrO_2$ | $SO_3$ | $V_2O_5$ | CuO | $P_2O_5$ | ZnO | PF |
| % | 52.2 | 36.00 | 1.91 | 0.68 | 0.49 | 0.25 | 0.09 | 0.05 | 0.04 | 0.02 | 0.02 | 0.01 | 8.24 |

**Table 4.4** Chemical analysis of soil drilling waste by fluorescence

| Soil drilling waste | | | | | | | | | | | | |
|---|---|---|---|---|---|---|---|---|---|---|---|---|
| Oxides | $SiO_2$ | CaO | $Al_2O_3$ | $Fe_2O_3$ | $K_2O$ | MgO | BaO | $SO_3$ | $TiO_2$ | SrO | $ZrO_2$ | CuO | ZnO | PF |
| % | 46.72 | 20.64 | 10.12 | 5.33 | 2.48 | 1.34 | 0.99 | 0.76 | 0.56 | 0.06 | 0.05 | 0.04 | 0.03 | 10.88 |

**Fig. 4.9** X-ray diffraction
pattern of soil

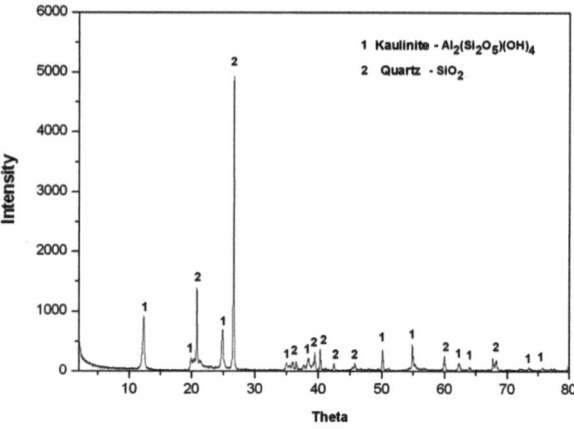

**Fig. 4.10** X-ray diffraction
pattern of soil drilling waste

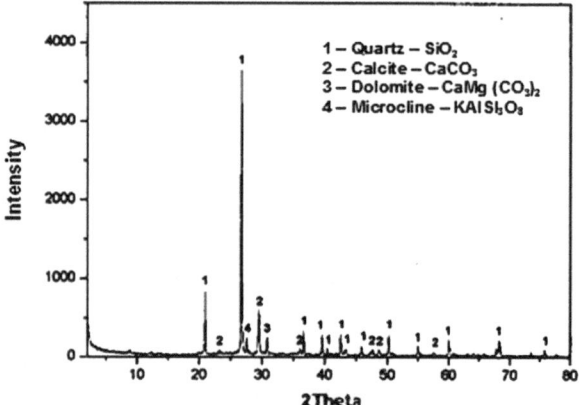

### 4.4.4   Thermal Analyses

Figure 4.11 illustrates the thermogravimetric curve of the soil drilling waste. Two
mass loss stages were observed, where the first loss occurs at temperature range of
100–500 °C and can be attributed to the dehydration of volatile organic compounds
from hydrocarbons incorporated into the soil drilling waste. The second stage
occurs at temperature range of 700–800 °C, which may be related to compound
stabilization. It was observed that at temperature range of 500–700 °C, there is heat
exchange, where quartz shifts from $\alpha$ phase to $\beta$ phase at around 570 °C.

### 4.4.5   Particle Size Analysis

Figure 4.12 shows the granulometric curve of the soil drilling waste, and it can be
seen that the material has 10 % of its volume in particle size lower than 3.55 μm in

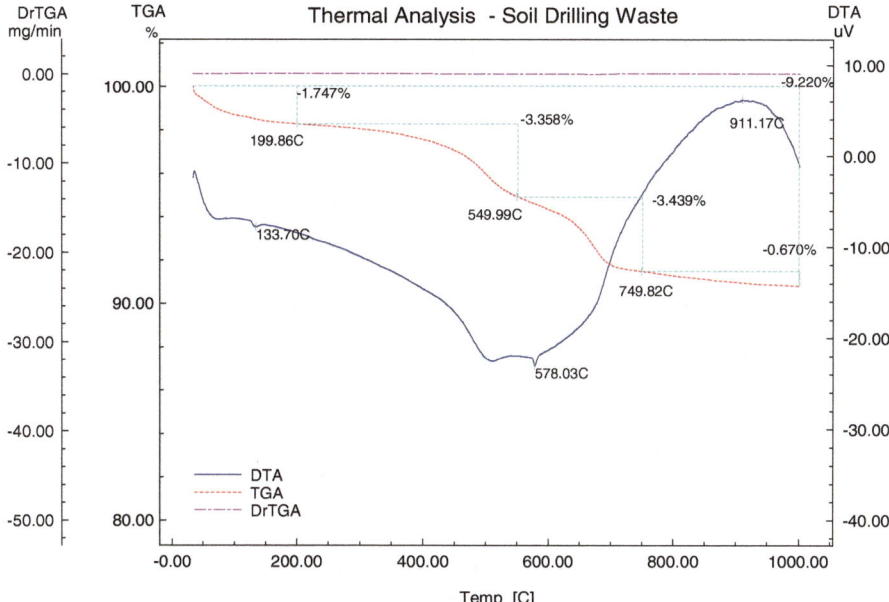

**Fig. 4.11** Thermogravimetric analysis of the soil drilling waste

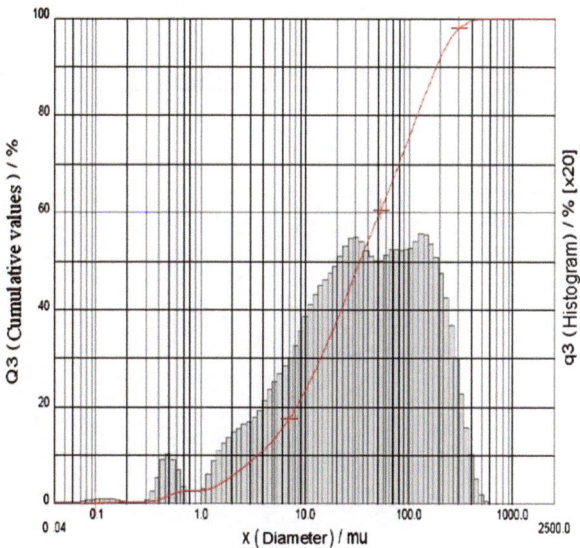

**Fig. 4.12** Particle size distribution of the soil drilling waste

which 50 % of particles have size smaller than 33.34 μm, and 90 % of particles are smaller than 181.08 μm, being classified as bimodal for presenting wide particle size distribution with two peaks. The average particle diameter is 66.83 μm.

### 4.4.6  Scanning Electron Microscopy (SEM)

Figures 4.13 and 4.14 show micrographs of the soil drilling waste. It was observed that the material is heterogeneous, with various particle sizes, corroborating the particle size analysis of the material.

**Fig. 4.13** Micrographs of the soil drilling waste obtained by SEM with magnification of ×500

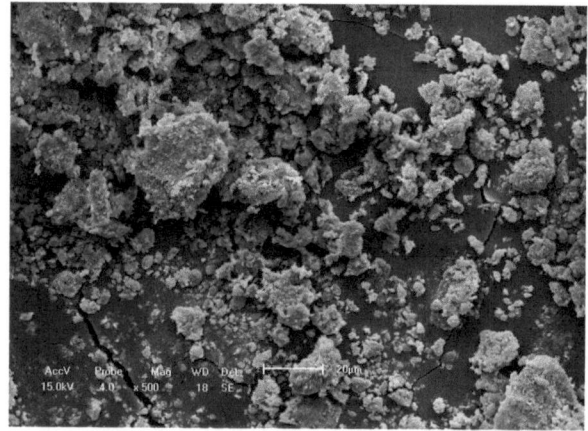

**Fig. 4.14** Micrographs of the soil drilling waste obtained by SEM with magnification of ×1000

## 4.5   Technological Tests

### 4.5.1   Water Absorption

The assay was performed according to NBR-8492, where bricks were dried in an oven until constant mass, removed, and weighed. Bricks were then immersed for 24 h in water, being removed for reweighing and calculation of the respective water absorbance values. It was found that all features met NBR 8492, which establishes maximum water absorption of 20 %. A decrease in water absorption values was observed as larger amounts of soil were added (Table 4.5). The greater the amount of waste, the lower the water absorption. It was also observed that the waste-cement mixture showed satisfactory absorption value.

### 4.5.2   Mass Loss by Immersion

The guidelines by ME 26—IPT/BNH have been followed—mass loss determination by wetting and drying of soil-cement bricks—test method. As shown in Table 4.6, bricks showed mass loss within standards, which establishes mass loss for soil-cement brick of up to 5 %. It was observed that bricks with maximum waste levels showed lower mass loss and as the content of waste is reduced in the formulation, bricks showed increased mass loss.

**Table 4.5**  Water absorption test of molded specimens

| Formulation | Water absorption (%) | Standard deviation |
|---|---|---|
| Soil + 14 % cement | 16.2 | ±0.3 |
| Drilling waste + 14 % cement | 14.0 | ±0.3 |
| Soil + 14 % cement + 80 % waste | 14.7 | ±0.3 |
| Soil + 14 % cement + 70 % waste | 14.6 | ±0.4 |
| Soil + 14 % cement + 60 % waste | 15.7 | ±0.3 |
| Soil + 14 % cement + 50 % waste | 15.9 | ±0.3 |

**Table 4.6**  Mass loss by immersion

| Formulation | Mass loss by immersion (%) | Standard deviation |
|---|---|---|
| Soil + 14 % cement | 4.8 | ±0.11 |
| Drilling waste + 14 % cement | 3.6 | ±0.11 |
| Soil + 14 % cement + 80 % waste | 3.9 | ±0.07 |
| Soil + 14 % cement + 70 % waste | 4.1 | ±0.11 |
| Soil + 14 % cement + 60 % waste | 4.1 | ±0.07 |
| Soil + 14 % cement + 50 % waste | 4.5 | ±0.11 |

**Table 4.7** Compressive strength of molded specimens

| Formulation | Compressive strength (MPa) | Standard deviation |
|---|---|---|
| Soil + 14 % cement | 4.34 | ±0.04 |
| SDW + 14 % cement | 6.21 | ±0.02 |
| Soil + 14 % cement + 80 % SDW | 3.87 | ±0.06 |
| Soil + 14 % cement + 70 % SDW | 3.78 | ±0.02 |
| Soil + 14 % cement + 60 % SDW | 3.92 | ±0.04 |
| Soil + 14 % cement + 50 % SDW | 4.06 | ±0.05 |

### 4.5.3   Compressive Strength

Table 4.7 shows the dosage followed by the average strength as a function of the amount of waste. The total soil replacement for SDW provides improved resistance of the material. The incorporation of SDW in the formulation Soil + 14 % cement causes a decrease of the strength. The increase in waste-cement brick strength can be justified by the presence of a cementations material in the soil drilling waste, which fully met the Brazilian standard, with no need to add soil to correct formulation.

All results previously described confirm that soil-cement bricks can be produced with addition of gravel from oil well drilling with properties and values set by Brazilian standards for soil-cement bricks. The results also showed that there is no need for correction of the mixture by clayey soil, as the formulation composed of pure gravel with cement also met Brazilian standards. The use of this waste without the addition of soil allowed greater use of this residue in formulations and therefore greater environment relief.

## References

Barba A et al (1997) Raw materials for the manufacture of ceramic tiles, 2nd edn. Espanha, Institute of Ceramic Technology—ICT

Brazilian Association of Technical Standards—ABNT (1986) NBR 7182: Solo—compression test. Test method. Rio de Janeiro, RJ, 10 p

Brazilian Association of Technical Standards—ABNT (1989) NBR 10833: Massive brick manufacture and soil-cement hollow block with use of hydraulic press. Procedure. Rio de Janeiro, RJ, 3 p

Calmon JL et al (1998) Utilization of granite cutting waste in the production of soil cement bricks. In: National meeting of built environment technology and quality in construction process, Florianópolis

Ferraz RL et al (2001) Comparative study of some methods of soil-cement mixtures dosages. State University of Maringa

Ferraz RC, Faleiro ET, Freire WJ (2005) Physical and mechanical performance of clay soil stabilized with lime and sodium silicate order application in rural buildings. J Tropical Agric 35 (3):191–198

Lopes WGR (2002) The hand of mud in Brazil. I SIACOT—I Ibero-American Seminar of earth construction, Anais, 16–18 Sept 2002, Salvador, BA, Brazil

Marques SKJ (2010) Study of gravel incorporation of oil well drilling formulations for soil-cement bricks. Dissertation (Master), Federal University of Rio Grande do Norte, Natal, RN

Mehta PK, Monteiro PJM (1994) Concrete: structure, properties and materials. Pini, São Paulo

Vargas M (1977) Introduction to soil mechanics. McGraw-Hill, São Paulo

# Chapter 5
# Using Oil Drilling Waste and Sugarcane Bags Ash in Soil-Cement Formulations

**Abstract** This chapter describes the properties of soil-cement bricks obtained with the addition of waste from the drilling of oil wells and sugarcane bagasse ash (SBA).

**Keywords** Sugarcane bagasse ash · Soil-cement bricks · Soil drilling waste · Mechanical and physical properties

## 5.1 Preparation of Specimens

Specimens were prepared analogously to soil-cement bricks with drilling waste described in item in Sect. 4.1. Table 5.1 shows the formulations studied in this work.

The constituents of specimens, soil, soil drilling waste (SDW), cement, and sugarcane bagasse ash (SBA) were separated after weighing in order to be mixed to the composition of masses and subsequent conformation. The mixture of mortar components was manually performed, placing in a cement container in order to manually crush small clods present in the material. Subsequently, soil was added (in the formulations that had this constituent), and then the waste with manual homogenization so that the final product had uniform color. After this step, drinking water was gradually added and new homogenization was performed. The mixture was transferred to a container where homogenization was performed to press 1 (Fig. 5.1) for solid bricks and press 2 (Fig. 5.2) for hollow brick. The mold of presses gives shape to bricks, either solid or hollow. After pressing, the machine ejects the molded element that is removed from the press and placed on tables to start the curing process on a flat surface. After 6 h from molding and during the first 7 days, the elements were kept moist by successive wettings (every 02 h) with spray to ensure adequate curing (Figs. 5.3 and 5.4).

© The Author(s) 2016
W. Acchar and S.K.J. Marques, *Ecological Soil-Cement Bricks
from Waste Materials*, SpringerBriefs in Applied Sciences
and Technology, DOI 10.1007/978-3-319-28920-5_5

**Table 5.1** Formulations investigated

| Formulation |
| --- |
| Soil-cement |
| Waste + 14 % Cement |
| Waste + 12 % Cement + 2 % Ash |
| Waste + 10 % Cement + 4 % Ash |
| Waste + 8 % Cement + 6 % Ash |

**Fig. 5.1** Press used for molding solid bricks

**Fig. 5.2** Press used for molding hollow bricks

**Fig. 5.3**  Solid bricks in the curing process

After the curing process, bricks were separated for each technological assay (mass loss by immersion, water absorption, and compressive strength).

With the aim of improving the absorption results and enabling the use of the material without the need for application of roughcast and plaster, glaze was applied (Fig. 5.5) and again absorption test in glazed specimens was carried out.

## 5.2   Physical Characterization of Sugarcane Bagasse Ash (SBA)

### 5.2.1   Particle Size Analysis

Figure 5.6 shows the particle size distribution chart of SBA held by laser particle size test. It was observed that the ash showed particle size distribution with particle diameter ranging from 0.3 to 25.0 μm, and of these, 50 % of the volume corresponded to particle diameter lower than 7.0 μm, classifying SBA as unimodal material for having higher concentration of fine particles in its distribution. The average particle diameter was 6.91 μm.

**Fig. 5.4** Hollow bricks in the curing process

## 5.2.2   X-ray Diffraction

Figure 5.7 shows the XRD pattern of SBA, which identified the following phases: tridymite ($SiO_2$), cristobalite ($SiO_2$), silicon oxide ($SiO_2$), calcium oxide (CaO), aluminum oxide ($Al_2O_3$), and iron oxide ($Fe_2O_3$). Cristobalite and tridymite phases are crystalline forms of silica, commonly found at high temperatures (metastable phase).

## 5.2.3   Micrograph of Sugarcane Bagasse Ash (SBA)

Figures 5.8 and 5.9 show that SBA particles have round shape and are formed of particles of varying sizes and shapes with fibrous aspect and overlapped layers with

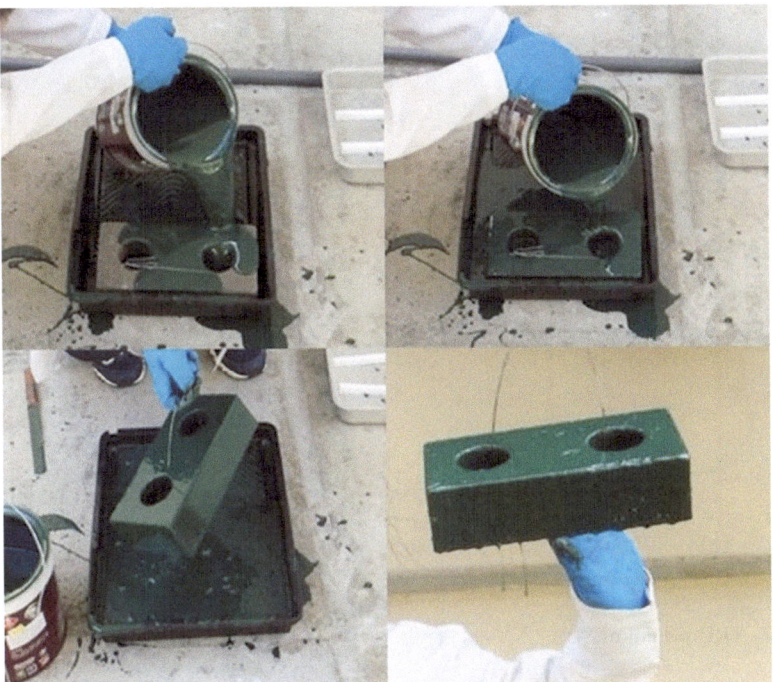

**Fig. 5.5** Hollow bricks after glaze application

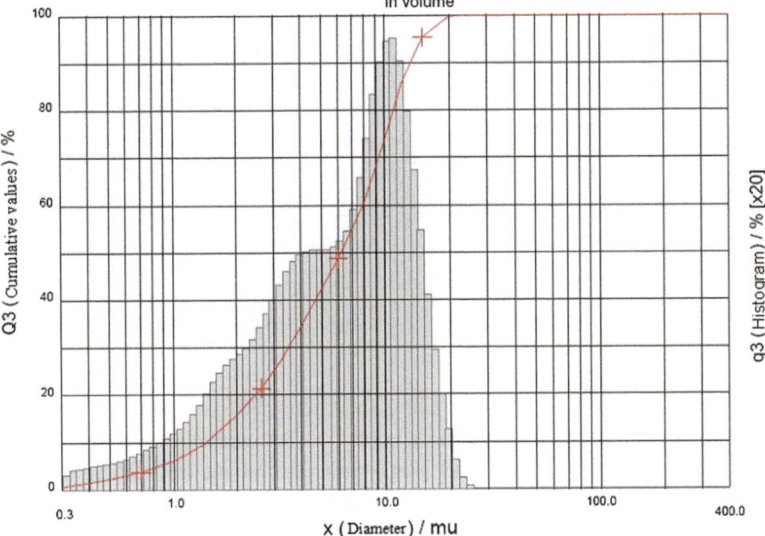

**Fig. 5.6** Particle size distribution of SBA

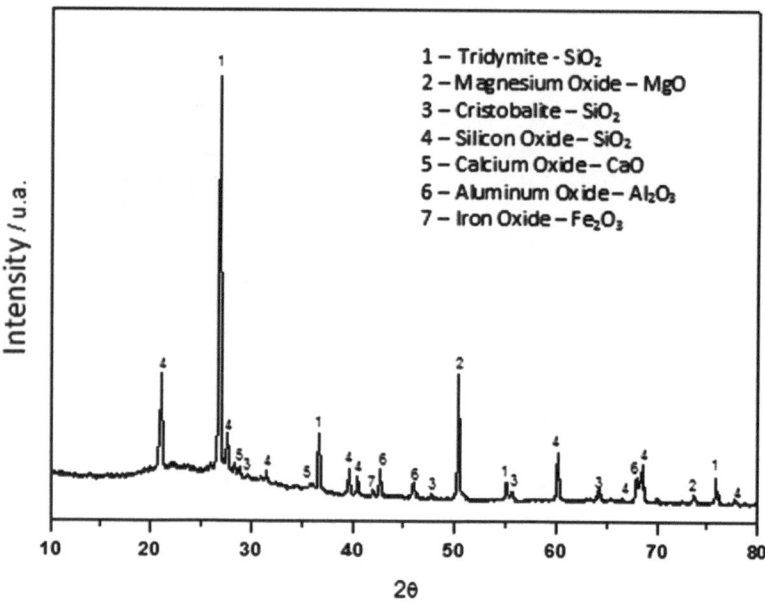

**Fig. 5.7** XRD pattern of SBA

**Fig. 5.8** Micrographs of
SBA obtained by SEM with
magnification of ×500

small particles adhered to the surface. Cordeiro (2006) observed that the morphology of ash particles produced at 600 °C is characterized by fibrous and microporous elements with varying sizes and phosphate and silica as main elements. SBA has high percentage of fine particles, thus enabling better homogeneity when associated with the SDW and high content of fine particles that will contribute to higher packing of particles.

**Fig. 5.9** Micrographs of
SBA obtained by SEM with
magnification of ×1000

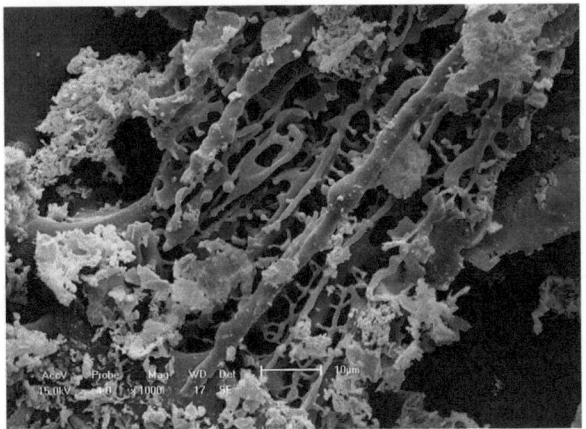

### 5.2.4   Thermal Analysis of Sugarcane Bagasse Ash

In Fig. 5.10, the thermogravimetric curve of SBA shows a minimum mass loss
around 1.6 %, which can be justified by the loss of moisture. After this small event,
it was observed that the ash had thermal stability, attributed to the burning process
to feed boilers for energy cogeneration. The ash DTA corroborated the results found
in X-ray diffraction for the presence of quartz fragments in the sample. It was
observed that in the temperature range from 400 to 500 °C, there is heat exchange,
where quartz shifts from α phase to β phase. The minimum mass loss value is close
to those found by Martirena Hernández et al. (2000), Massaza (1998) and
Freitas (2005).

### 5.2.5   Thermal Analysis of the Drilling Waste Composition
          and SBA

Figure 5.11 shows the thermogravimetric curve of the SDW and SBA (2 wt%
SBA). Two events were observed, where the first occurs with mass loss in the
temperature range of 200–750 °C, which can be attributed to the dehydration of
organic compounds, and the second event occurred above 750 °C, showing compound stability. Considering both events, the result is consistent with results of
thermal analysis of each material, where SDW showed mass loss around 9 % and
SBA around 1 %. The analysis of the matrix showed mass loss around 10 %.

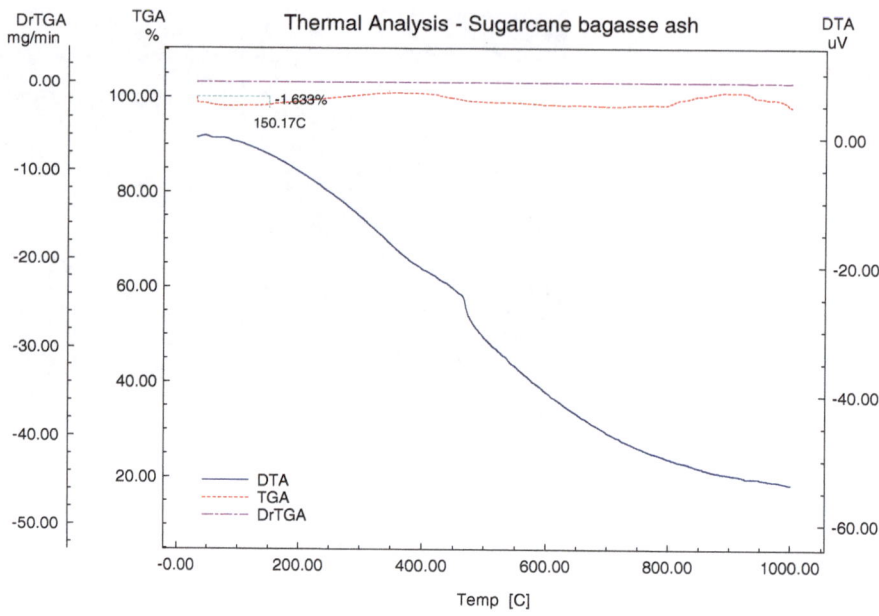

**Fig. 5.10**  Thermal analysis of SBA

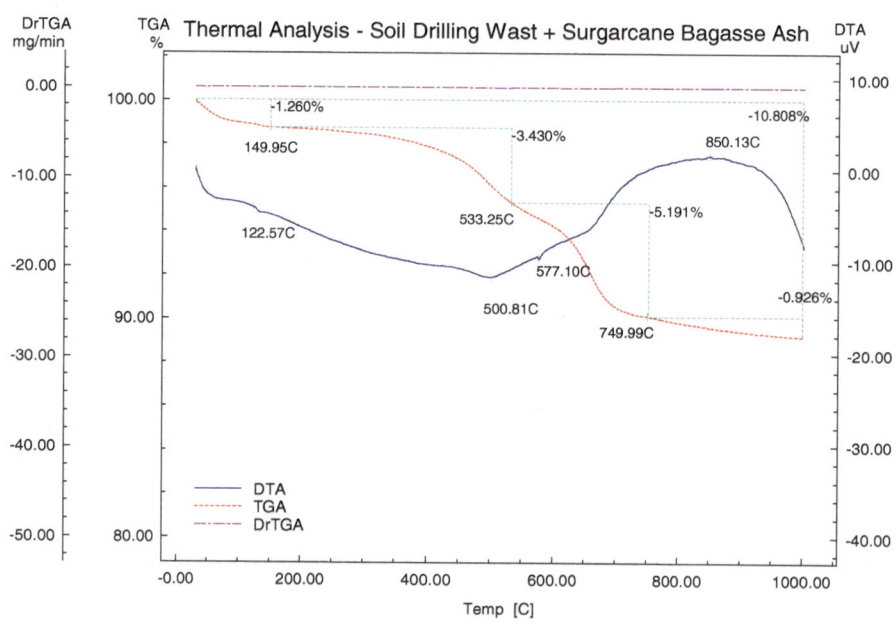

**Fig. 5.11**  Thermal analysis of soil drilling waste and SBA composition

**Table 5.2**   Chemical analysis of SBA by fluorescence

| Oxides | $SiO_2$ | $Al_2O_3$ | $Fe_2O_3$ | $TiO_2$ | MgO | $P_2O_5$ | $Na_2O$ | $SO_3$ | CaO | Others |
|--------|---------|-----------|-----------|---------|-----|----------|---------|--------|-----|--------|
| %      | 44.01   | 8.36      | 1.72      | 0.69    | 4.74| 3.49     | 0.14    | 4.71   | 6.84| 1.83   |

## 5.2.6   Chemical Analysis of SBA by Fluorescence

According to Table 5.2, $SiO_2$ is the predominant compound in SBA, with levels of about 44 % of the mass of samples, followed by $Al_2O_3$ with levels of about 9 %. $Fe_2O_3$, $TiO_2$, MgO, $P_2O_5$, $SO_3$, and CaO represent about 23 % of the ash mass. According to Cordeiro (2006), the prevalence of silica is probably due to silicon adsorbed from the soil by sugarcane plant roots in the form of monosilicic acid ($H_4SiO_4$). The author reports that it can also be due to the presence of sand impregnated in sugarcane plant roots. The percentage of silica obtained was below values found in literature (Cordeiro 2006). This can be explained by the method used to harvest sugarcane. According to literature, most of the silicas found in SBA may result from the presence of sand impregnated in sugarcane plant roots at the time of mechanical harvesting.

## 5.3   Technological Tests

### 5.3.1   Water Absorption in Solid Bricks

It was found that all traces met NBR 8492 standard (2012), which establishes maximum absorption of 20 %. There is a decrease in absorption values as ash is added (Table 5.3).

### 5.3.2   Water Absorption of Hollow Bricks

Test was performed according to 10836 NBR, which establishes that bricks must present average water absorption values equal to or smaller than 20 % and

**Table 5.3**   Water absorption of solid bricks

| Dosage—solid brick | Water absorption (%) | Standard deviation |
|--------------------|----------------------|--------------------|
| Soil-cement | 16.2 | ±0.03 |
| Waste + 14 % Cement | 16.2 | ±0.03 |
| Waste + 12 % Cement + 2 % Ash | 14.0 | ±0.03 |
| Waste + 10 % Cement + 4 % Ash | 14.7 | ±0.03 |
| Waste + 8 % Cement + 6 % Ash | 15.7 | ±0.03 |

**Table 5.4** Water absorption of hollow bricks

| Formulation | Water absorption (%) | SD |
|---|---|---|
| Soil-cement | 17.6 | ±0.03 |
| Waste + 14 % Cement | 16.5 | ±0.03 |
| Waste + 12 % Cement + 2 % Ash | 15.7 | ±0.03 |
| Waste + 10 % Cement + 4 % Ash | 15.3 | ±0.03 |
| Waste + 8 % Cement + 6 % Ash | 15.9 | ±0.03 |

*SD* Standard deviation

individual values equal to or smaller than 2 % at 28 days of age. All formulations for hollow bricks met the standard and as ash was added to formulations, water absorption values showed a decrease up to ash content of 4 %. Formulation with 6 % showed a small increase in absorption, as shown in Table 5.4.

### 5.3.3  Mass Loss by Immersion of Solid Bricks

The guidelines of ME 26—IPT/BNH—mass loss determination by wetting and drying of soil-cement bricks—Test method, have been followed. As shown in Table 5.5, bricks showed mass loss within standards specified by legislation, which establishes mass loss for soil-cement bricks of 5 %. It was observed that bricks molded with maximum waste content obtained lower mass loss and as the waste content is decreased in the formulation, bricks showed greater mass loss.

### 5.3.4  Mass Loss by Immersion of Hollow Bricks

The guidelines of ME 26—IPT/BNH—mass loss determination by wetting and drying of soil-cement bricks—Test method, have been followed. As shown in Table 5.6, bricks showed mass loss within standards specified by legislation, which establishes mass loss for soil-cement bricks of 5 %. Even with the same formulation of solid bricks, hollow bricks showed a small increase in mass loss.

**Table 5.5** Mass loss by immersion of solid bricks

| Formulation | Mass loss (%) | SD |
|---|---|---|
| Soil-cement | 2.8 | ±0.08 |
| Waste + 14 % Cement | 2.6 | ±0.07 |
| Waste + 12 % Cement + 2 % Ash | 2.5 | ±0.07 |
| Waste + 10 % Cement + 4 % Ash | 2.7 | ±0.07 |
| Waste + 8 % Cement + 6 % Ash | 3.1 | ±0.07 |

*SD* Standard deviation

| Formulation | Mass loss (%) | SD |
|---|---|---|
| Soil-cement | 2.9 | ±0.08 |
| Waste + 14 % Cement | 2.6 | ±0.07 |
| Waste + 12 % Cement + 2 % Ash | 2.9 | ±0.07 |
| Waste + 10 % Cement + 4 % Ash | 3.1 | ±0.07 |
| Waste + 8 % Cement + 6 % Ash | 3.5 | ±0.07 |

**Table 5.6** Mass loss by immersion of hollow bricks

*SD* Standard deviation

## 5.3.5  Compressive Strength of Solid Bricks

Table 5.7 shows the composition and average compressive strength as a function of the amount of waste and ash. The increased strength of waste cement bricks can be justified by the presence of a cementations material in the drilling waste. In formulations where cement has been partially replaced by ash, an increase in strength is observed, which can be attributed to the presence of silicon oxide that along with aluminum oxide in the presence of water will give pozzolanic effect to the material. The particle size distribution of the SDW and ash with higher content of fine particles can also be considered, which contribute to a better packing of particles resulting in a reduction in voids. It was observed that the average strength of bricks, with and without addition of ash, in trial carried out after 14 days of curing of specimens, has met the requirements of NBR 8492 (ABNT 1984). This standard establishes average values greater than or equal to 2.0 MPa after seven days. All compositions met the minimum requirements of Brazilian standards.

## 5.3.6  Compressive Strength of Hollow Bricks

Table 5.8 shows the dosage followed by average strength as a function of the amount of SDW and ash. Even with the same formulation of solid bricks, hollow brick showed lower strength, which can be correlated with the specific surface of bricks, where the two holes in the material bring about higher possibility of cracks and fissures. Even presenting lower strength, all formulations met the standard that establishes compressive strength of 2.0 MPa.

**Table 5.7** Compressive strength of solid bricks

| Formulation | Compressive strength (MPa) | Standard deviation |
|---|---|---|
| Soil-cement | 4.34 | ±0.03 |
| Waste + 14 % Cement | 6.21 | ±0.03 |
| Waste + 12 % Cement + 2 % Ash | 7.94 | ±0.03 |
| Waste + 10 % Cement + 4 % Ash | 5.33 | ±0.03 |
| Waste + 8 % Cement + 6 % Ash | 3.64 | ±0.03 |

**Table 5.8** Compressive strength of hollow bricks

| Formulation | Compressive strength (MPa) | Standard deviation |
|---|---|---|
| Soil-cement | 4.14 | ±0.03 |
| Waste + 14 % Cement | 5.19 | ±0.03 |
| Waste + 12 % Cement + 2 % Ash | 5.60 | ±0.03 |
| Waste + 10 % Cement + 4 % Ash | 4.15 | ±0.03 |
| Waste + 8 % Cement + 6 % Ash | 3.10 | ±0.03 |

**Table 5.9** Water absorption after glaze application in solid bricks

| Formulation | Water absorption (%) | SD |
|---|---|---|
| Soil-cement | 14.2 | ±0.03 |
| Waste + 14 % Cement | 14.2 | ±0.03 |
| Waste + 12 % Cement + 2 % Ash | 13.0 | ±0.03 |
| Waste + 10 % Cement + 4 % Ash | 13.7 | ±0.03 |
| Waste + 8 % Cement + 6 % Ash | 14.7 | ±0.03 |

### 5.3.7   Water Absorption After Enamel of Solid Bricks

After water absorption evaluation, even though all formulations have met standards, glaze was applied on bricks (solid and hollow) in order to further decrease the absorption and improve their finishing to be used in the apparent form, eliminating the need for use of roughcast and plaster (Fig. 5.5). Table 5.9 shows the absorption of solid bricks after glaze application, in which a decrease in absorption in all formulations in relation to unglazed formulation could be observed.

### 5.3.8   Water Absorption After Glaze Application in Hollow Bricks

Table 5.10 shows the decrease in the water absorption of hollow bricks after glaze application, which may be justified by the thin film that glaze forms on the material, preventing water to infiltrate more easily into the material.

**Table 5.10** Water absorption after glaze application in hollow bricks

| Formulation | Water absorption (%) | SD |
|---|---|---|
| Soil-cement | 16.6 | ±0.03 |
| Waste + 14 % Cement | 16.5 | ±0.03 |
| Waste + 12 % Cement + 2 % Ash | 14.7 | ±0.03 |
| Waste + 10 % Cement + 4 % Ash | 14.3 | ±0.03 |
| Waste + 8 % Cement + 6 % Ash | 15.9 | ±0.03 |

## 5.4   Analysis of the Fracture Surface

Figures 5.12, 5.13, and 5.14 present the micrographs of fracture surfaces of specimens concerning the best result obtained by the formulation with ternary combination, waste cement, with the percentages by weight of 12 % of cement and 2 % of
ash. The images reinforce results previously obtained, namely, the development of
good physical homogeneity, especially for good particle size distribution, which
can be evidenced by the microstructural uniformity evidenced by micrographs.

The material microstructure and the type and distribution of constituent phases
play a key role between the process of material formation and its properties
(Scrivener et al. 2004). The micrographs show general characteristics associated
with cement hydration products. Needle-like crystals composed of hydrated tri-
calcium sulphoaluminate and calcium sulphoaluminate hydrate even at early ages
were found. Micrographs have shown the evolution of fibrous crystals of calcium
silicate hydrate, resembling leaves and large hexagonal prismatic crystals that may
be typical of portlandite (calcium hydroxide) and calcium aluminate hydrate
(Figs. 5.13 and 5.14).

It was also observed that bricks are characterized by a compact structure. The
pore structure of percentages evaluated in relation to reference does not show
marked morphological changes. This may suggest that the presence of SCA did not
influence the formation of hydration products.

**Fig. 5.12** Micrographs
obtained by SEM of the
fracture surface, with
magnification of ×3000

**Fig. 5.13** Micrographs of the fracture surface obtained by SEM, with magnification of ×8000

**Fig. 5.14** Micrographs of the fracture surface obtained by SEM, with magnification of ×500

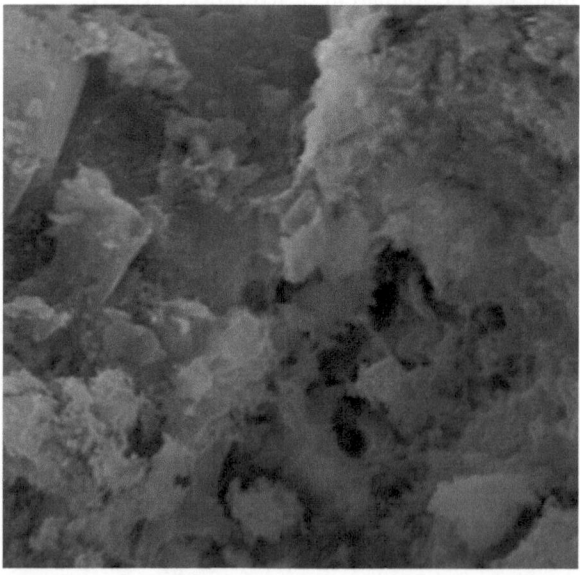

## 5.5  Correlations Model Between Mechanical Properties and Chemical Reactions

Calcium silicates are almost solely responsible for the mechanical characteristics measured in the cement paste. These silicates—$C_3S$ and $C_2S$—make up about 65–85 % of cement weight (Type I) and upon hydration, form the C–S–H gel, the major component of hydrated cement paste (Mehta and Monteiro 1994). In order to simplify the cement hydration study, it is usual to replace the complex water/cement system by the water/silicate system. The replacement is acceptable because silicates are the most effective constituents for being present in greater proportion. The morphology of hydrosilicates presents fibrous particles with dimensions of a few microns that cover grains of anhydrous silicates. In general, the material is little crystalline and forms a porous solid that exhibits characteristics of a rigid gel (Neville 1997). Whereas the chemical composition of hydrated calcium silicates varies with water/cement ratio, temperature and hydration age, it has become common to refer to these hydrosilicates simply as C–S–H, but the notation does not imply a fixed composition. In the case of complete hydration, the approximate composition of the material corresponds to $C_3S_2H_3$, which is used for stoichiometric calculations. During hydration, calcium silicate hydrate microcrystals C–S–H, as small filaments of a felt, arise on the cement surface and crystallize. At the end of hydration, macrocrystals form a total specific surface of around 200 $m^2/g$ (Powers 1968). Given the nature of the crystallization surface, macrocrystals adhere to each other and intertwine, also adhering to aggregated crystals, thereby forming a solid structure. Cement components rich in CaO strongly react with water producing C–S–H gel and calcium hydroxide ($Ca(OH)_2$), and releasing heat. The heat generated in the hydration is also indirectly responsible for the mechanical strength of the paste, as it can generate thermal shrinking microcracks.

The main reactions responsible for mechanical strength are shown below (Mehta and Monteiro 1994):

$$C_3S + H_2O \rightarrow C - S - H \text{ gel} + Ca(OH)_2 + 120 \text{ cal}/g \text{ of } C_3S \qquad (5.1)$$

Equation 5.2 provides high initial strength and strong hydration heat release (about 80 % in 10 days):

$$C_2S + H_2O \rightarrow C - S - H \text{ gel} + Ca(OH)_2 + 60 \text{ cal}/g \text{ of } C_2S \qquad (5.2)$$

Equation 5.3 provides slow and steady strength development and low hydration heat (about 80 % after 100 days). Since the formation of calcium silicate hydrate is one of the reasons for the paste hardening, it was possible to observe a significant increase in the strength of solid and hollow bricks. Even with the partial replacement of cement by SBA, no decrease in strength was observed because calcium oxide and silica components present in the drill SDW and SBA enabled the formation of calcium silicate hydrate. The reactions involved in the process of formation of calcium silicate hydrate are represented as follows:

$$2Ca_3SiO_5 + 7H_2O \rightarrow 3CaO \cdot 2SiO_2 \cdot 4H_2O + 3Ca(OH_2) + 173.6\,kJ \qquad (5.3)$$

In addition to the formation of $Ca(OH_2)$ through hydration reactions of the cement components (Eqs. 5.1 and 5.2), there is also the reaction of CaO (Eq. 5.4) from the ash, according to DRX:

$$CaO + H_2O \rightarrow Ca(OH)_2 + 275\,cal/g\,CaO \qquad (5.4)$$

CaO formed in these reactions may react with $SiO_2$ present both in drilling waste as in SBA, as pozzolanic reaction (Eq. 5.5). The greatest effect for the occurrence of this reaction with the addition of the SBA, which has metastable phases of silica, such as cristobalite and tridymite, possibly has some amorphous silica phases with a higher reactivity:

$$3Ca(OH)_2 + 2SiO_2 \rightarrow 3CaO \cdot 2SiO_2 \cdot 3H_2O \qquad (5.5)$$

Increased strength can be associated with increased formation of C–S–H due to the pozzolanic reaction with the addition of ash. The pozzolanic reaction leads to a reduction of free $Ca(OH)_2$ associated to a refining of pore size, which may be contributing to reduce both the water absorption and the mechanical strength.

The lack of increase in mechanical strength with higher ash contents can be associated with the lack of increase in the formation of C–S–H by the pozzolanic reaction, since 2 % of SBA can provide the stoichiometric quantities necessary for the occurrence of the reaction, while the additional amount of ash is acting only as filler (load) in the mass without effective contribution to the mechanical properties.

## 5.6  Cost of Solid and Hollow Bricks

Budgeting is the verification of costs, and is a part of a strategic financial plan including the forecast of future revenues and expenses for the administration of a certain exercise (time period). Aiming at estimating the cost of solid and hollow bricks, a survey was conducted based on the cement used, keeping in mind that the wastes used are disposed in the environment and do not yet have any commercial value. After calculating the amount of cement and verifying the cost of a sack of cement of 50 kg, a table with the cost was elaborated. Table 5.11 shows that the cost of bricks ranges from 0.03 to 0.07 cents depending on the percentage of incorporated cement. If we consider that one thousand conventional bricks cost, on average, R$ 390.00, the brick that showed the best mechanical results was formulated with 2 % ash, one thousands of these brick would cost R$ 50.00. In addition to the far lower cost compared to conventional brick, the cost of roughcast and plaster can be excluded, which will further minimize the total costs of the work.

**Table 5.11** Table showing the cost of solid and hollow bricks

| Formulation | Values (R$) |
| --- | --- |
| Soil-cement | 0.07 |
| Waste + 14 % Cement | 0.07 |
| Waste + 12 % Cement + 2 % Ash | 0.05 |
| Waste + 10 % Cement + 4 % Ash | 0.04 |
| Waste + 8 % Cement + 6 % Ash | 0.03 |

# References

Brazilian Association of Technical Standards—ABNT (1984) NBR 8492: Massive soil-cement bricks—determination of compressive strength and Moorhen absorption. Test method. Rio de Janeiro, RJ, 5 p

Cordeiro GC (2006) Use of ultrafine ash from bagasse from sugarcane and rice husk as mineral additives. Thesis (Doctorate in Civil Engineering)—Rio de Janeiro—RJ, Federal University of Rio de Janeiro, UFRJ, 445 p

Freitas ES (2005) Characterization bagasse ash from cane sugar of Campos dos Goytacazes municipality for use in construction. Dissertation (Master in Civil Engineering)—Campos dos Goytacazes—RJ, State University of North Fluminense UENF, 81 p

Martinera Hernández JF, Betancourt Rodríguez S, Middendorf B, Rubio A, Martínez Fernández L, Machado López I, González López R (2000) Pozzolanic properties of the waste sugar industry (part one). Build Mater 50:71–78

Massazza F (1998) Pozzolana and pozzolanic cements. In: Hewlett PC (ed) Lea's chemistry of cement and concrete, 4th edn. Arnold Publishers, London

Mehta PK, Monteiro PJM (1994) Concrete: structure, properties and materials. Pini, São Paulo

Neville AM (1997) Concrete properties. Trad. Salvador E. Giammusso. 2nd edn. Pini, Sao Paulo

Powers C (1968) The properties of fresh concrete, New York

Scrivener KL et al (2004) Quantitative study of Portland cement hydration by X-ray diffraction/Rietveld analysis and independent methods. Cem Concr Res 34(9):1541–1547

# Chapter 6
# Conclusions

**Abstract** This chapter shows the main conclusions about the use of both waste materials.

**Keywords** Soil drilling waste · Sugarcane bagasse ash

After analysis and discussion of results obtained in tests, following could be concluded:

**Soil drilling waste from the petroleum industry (SDW)**

The incorporation of gravel from oil wells in the manufacture of soil-cement bricks is technically feasible and can be an alternative to the use of this waste, reducing environmental impact caused by the disposal of this waste in landfills. The formulation with the highest strength was brick made with 100 % waste, showing compression strength of 6.21 MPa and lower water absorption, thus confirming the potential of gravel waste and the lack of need for addition of clayey soil to correct the plasticity of the mixture to be compacted. Bricks produced by incorporating gravel waste under study had improved mechanical properties and all of them met the minimum requirements of Brazilian standards;

**Sugarcane bagasse ash (SBA) and Soil drilling waste (SDW)**

The incorporation of soil drilling waste (SDW) and ash from sugarcane bagasse (SGA) burning in the manufacture of waste-cement bricks proved to be technically feasible and can be configured as an alternative to the recovery of waste.

Bricks produced by incorporating both wastes (SDW and SGA) had improved mechanical properties and all of them met the minimum requirements of Brazilian standards; it is therefore an environmentally friendly practice as it can contribute toward transforming materials that would otherwise be discarded in nature into important source of raw material.

© The Author(s) 2016
W. Acchar and S.K.J. Marques, *Ecological Soil-Cement Bricks*
*from Waste Materials*, SpringerBriefs in Applied Sciences
and Technology, DOI 10.1007/978-3-319-28920-5_6

The incorporation of both residues into soil-cement bricks contributes not only to reduce impacts to the environment, but also to reduce the costs involved in the production of bricks. As shown in previous studies, incorporation of both residues as raw materials in the production of soil-cement bricks lead to a reduction in the total production cost compared to conventional bricks using cement only.